Dam Hydraulics

Wiley Series in Water Resources Engineering

The 1992 'International Conference on Water and the Environment: Issues for the 21st Century' served as a timely reminder that fresh water is a limited resource which has an economic value. Its conservation and more effective use becomes a prerequisite for sustainable human development.

With this in mind, the aim of this series is to provide technologists engaged in water resources development with modern texts on various key aspects of this very broad discipline.

Professor J. R. Rydzewski
Irrigation Engineering
Civil Engineering Department
University of Southampton
Highfield
SOUTHAMPTON
S09 5NH
UK

Design of Diversion Weirs: Small Scale Irrigation in Hot Climates
Rozgar Baban

Unit Treatment Processes in Water and Wastewater Engineering
T. J. Casey

Water Wells: Implementation, Maintenance and Restoration
M. Detay

Dam Hydraulics
D. L. Vischer   W. H. Hager

Dam Hydraulics

D. L. Vischer

W. H. Hager
ETH-Zentrum, Zürich, Switzerland

JOHN WILEY & SONS
Chichester · New York · Weinheim · Brisbane · Singapore · Toronto

Copyright © 1998 by John Wiley & Sons Ltd,
Baffins Lane, Chichester,
West Sussex PO19 1UD, England

National 01243 779777
International (+44) 1243 779777

e-mail (for orders and customer service enquiries): cs-books@wiley.co.uk

Visit our Home Page on http://www.wiley.co.uk
or
http://www.wiley.com

Cover Illustration: Karakaya dam, Turkey (Courtesy of Elektrowatt Engineering, Zurich,
Switzerland)

Other Wiley Editorial Offices

John Wiley & Sons, Inc., 605 Third Avenue,
New York, NY 10158-0012, USA

Wiley-VCH Verlag GmbH, Pappelallee 3,
D-69469 Weinheim, Germany

Jacaranda Wiley Ltd, 33 Park Road, Milton,
Queensland 4064, Australia

John Wiley & Sons (Asia) Pte Ltd, 2 Clementi Loop #02-01,
Jin Xing Distripark, Singapore 0512

John Wiley & Sons (Canada) Ltd, 22 Worcester Road,
Rexdale, Ontario M9W 1L1, Canada

Library of Congress Cataloging-in-Publication Data

Vischer, D. (Daniel)
    Dam hydraulics/D. L. Vischer, W. H. Hager.
      p.   cm. — (Wiley series in water resources engineering)
    Includes bibliographical references (p.   ) and index.
    ISBN 0-471-97289-4 (pbk.)
    1. Dams.   2. Hydraulics.   I. Hager, Willi H.   II. Title.
  III. Series.
  TC540.V57 1997
  627'.8 — dc21   97-7325
                                                        CIP

British Library Cataloguing in Publication Data

A catalogue record for this book is available from the British Library

ISBN 0 471 97289 4

Typeset in 10/12 Times by Pure Tech (India) Ltd

Contents

Preface

Dam hydraulics relates to the hydraulic works that have to be under-taken when planning, constructing and refurbishing dams. It is of major significance for two reasons:

1. In the industrialized countries, there are currently thousands of dams that have to be refurbished. Renovations include the *spillways* that have to be upgraded to convey larger design discharges or to protect against abrasion and cavitation, the *bottom outlets* that have not been included in the original design or that do not allow appro-priate flushing of sediments, and the *intake structures* that have not been brought to the modern requirements regarding efficiency. More-over, questions relating to hydraulic problems such as reservoir sedi-mentation, impulse waves originating from slides or dambreak waves will have to be studied in the future.

2. In developing countries, a balance has to be set up between the natural runoff and the demands for drinking and irrigation water supply as well as for hydropower. Other problems relate to the increase of the minimum water levels in navigated rivers or to flood protection in river systems. In these countries, hundreds of dams are currently being planned or constructed, that have to be designed on a sound hydraulic basis.

Dam Hydraulics addresses the main topics of water flow in dam structures. The authors have aimed at stressing the hydraulic principles that govern the current design practice. The various phenomena are illustrated with appropriate figures. In addition, selected photographs of laboratory models and prototype structures have been included to both illustrate the actual flow features and to give impressions of the natural beauty of water flow in such structures. The authors are of the opinion that the aesthetics of flow phenomena are a definite point of motivation for both education and profession.

The readership of this book includes graduate students and profes-sional engineers involved in the design of dams; it can also give insights to environmental engineers, geologists, and other scientists who have to deal with dams in general. The main purpose of the book is thus to convey knowledge readily applicable to design. A number of selected references of mainly Anglo-Saxon, French, Italian and German sources is given at the end of each chapter. All symbols are explained in the text when appearing first. Detailed subject and author indices can be found at the end of the book.

We would like to acknowledge the reviews of Prof. Vijay P. Singh, Louisiana State University, Baton Rouge LA, and Prof. George Christodoulou, National Technical University of Athens, Greece during their visits at VAW. The assistance of Dr. Karin Schram, VAW, for preparing this book is appreciated.

Zurich
December 1996

Karakaya arch dam, Turkey (Courtesy of Italstrade S.p.A. Milano, Italy)

1

Introduction

1.1 WHAT IS DAM HYDRAULICS?

Dams are erected to create a reservoir, an hydraulic head or a water surface:

- A *reservoir* is used to coordinate water yield with water needs. The reservoir serves thus for the temporal storage of water. Typical are dams for drinking water supply, irrigation or hydropower. Usually, water is stored during flood periods and used during dry seasons. Dams are thus also effective structures for flood protection, as they store water and release it with a temporal lag.

- An *hydraulic head* increases the net pressure on a power plant. Also, a river can be improved for navigation by a tailwater creation.

- A *water surface* enables navigation and lake recreation.

Dam hydraulics considers therefore all hydraulic questions that relate to the construction, the management and the safety of dams. During construction, the river has to be diverted with channels, tunnels or culverts. A bottom outlet is imperative during the first filling of the reservoir. It assures a filling control and may later be incorporated for reservoir emptying. The emptying process is necessary during danger, revision or for reservoir flushings. For withdrawal of water a separate intake structure is arranged. To prevent overtopping of a dam, an overflow structure is integrated that is capable to divert even extra-ordinary floods without significant damage. Dam hydraulics is thus directed to the *hydraulic design* of the following items:

- diversion during construction,
- bottom outlets,
- intake structures, and
- overflow structures.

Particular hydraulic problems include vortex formation at intakes, air entrainment, cavitation and vibration, energy dissipation and erosion. Special topics of dam hydraulics that are also considered are:

- reservoir sedimentations,
- impulse waves due to slope instabilities, and
- dambreak waves.

1.2 DESCRIPTION OF DAM HYDRAULICS

The various parts of a dam can be described with selected typical structures. In the following three dams are described, namely the Karakaya arch dam, the Itaipu concrete dam and the La Grande 3 rockfill dam. Also an introduction relating to dam hydrology is discussed in the following sections.

Figure 1.1 refers to the *Karakaya dam* in Turkey. This multipurpose development on the Euphrates river is located between the Keban dam and the Ataturk dam, and is one of the largest schemes for the generation of energy and for irrigation purposes worldwide. The catchment area has some $80\,000\,km^2$ and the mean average discharge is $725\,m^3s^{-1}$. The total capacity of the reservoir is almost $10 \times 10^9\,m^3$ and its surface is $300\,km^2$ over a length of $166\,km$. The maximum height of the dam amounts to $173\,m$ and the crest length is $462\,m$. The design discharge is $17\,000\,m^3s^{-1}$ and the maximum discharge that can be discharged is $22\,000\,m^3s^{-1}$. The power generated is $1800\,MW$ and the annual production can be up to $7100\,GWh$. The head on the Francis turbines has an average of $150\,m$.

Karakaya is a concrete dam with ten gated overflow structures each $14\,m$ wide and discharging into three spillways concentrically arranged. The spillway terminates in a trajectory spillway that discharges the water back into the Euphrates river. Two streamlined

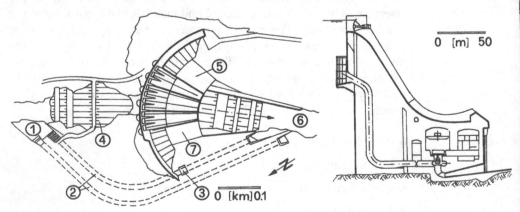

Figure 1.1 Karakaya dam (Turkey) (a) plan and (b) principal section with ① intake structure, ② 11.5 m diameter diversion tunnels, left tunnel used as bottom outlet, right tunnel closed after construction, ③ bottom outlet chambers, ④ upstream cofferdam, ⑤ service building, ⑥ outlet works, ⑦ power house (Zanon, 1988)

Figure 1.2 Karakaya spillway from the tailwater (*Water Power & Dam Construction* 41(7): 20)

wingwalls are provided upstream and laterally of the overflow weir and guidewalls incorporate three, four and three gate bays, respectively. The design head on the overflow structures is 13 m. Construction began in 1975 and completion was in 1988 (Stutz, et al., 1979).

Itaipu dam is the largest hydropower installation worldwide. It is located on the river Parana and owned by Brazil and Paraguay. Figure 1.3 shows plan and principal section of the dam. The Itaipu power scheme has a reservoir of 170 km long and up to 8 km wide. The

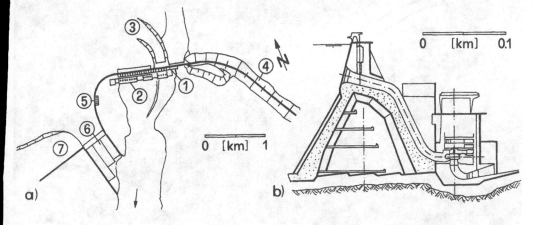

Figure 1.3 Itaipu dam (Brazil-Paraguay) (a) plan, (b) principal section with ① main dam, ② power-house, ③ diversion channel, ④ left earthfill/rockfill dam, ⑤ lateral intake, ⑥ spillway, ⑦ right earthfill dam

concrete dam is 2.6 km long and up to 196 m high. The earth dams are 5 km long. The spillway is designed for a design discharge of $62\,000\,\mathrm{m^3s^{-1}}$ with a return period of 10 000 years. The power house is nearly 1 km long and equipped with 18 8.5 m diameter Francis turbines of 700 MW each.

During the construction period work continued round the clock involving 35 000 persons. The dimensions of the schemes are gigantic: 64 million $\mathrm{m^3}$ of rock and earth have been removed, 12 million $\mathrm{m^3}$ of concrete and 500 000 tonnes of steel were used. In 1984, the first electric power was supplied to the city of Sao Paolo (Brazil). The environmental impact was considered significant: 10 000 farmers had to be resettled and $1350\,\mathrm{km^2}$ were flooded. A forest area of $230\,\mathrm{km^2}$ was provided along the shore of the reservoir to prevent erosion. The wildlife was rescued and transferred into reservation zones. In the reservoir, 125 species of fish have settled and a fishway on the neighboring Acaray river was installed. Figure 1.4 shows a view of the scheme.

The spillway design was described by De Moraes, et al. (1979). Three chutes were selected as a compromise between economy and operational flexibility. The design head of the spillway crest was 20 m and the maximum head was set to 23 m. In total 14 gates each 20 m × 20 m are provided and the maximum capacity of the spillway is $62\,000\,\mathrm{m^3s^{-1}}$.

Figure 1.4 Itaipu dam, aerial view (Courtesy of N. Pinto)

The pier at the left side of the spillway was elongated to reduce over-topping, and the other pier ends had a square end because some accidential overtopping over the intermediate walls is tolerated.

The total width of the three chutes including the guide walls is 350 m. This width is maintained constant over the entire chute. The invert of the chutes follows essentially the available sound rock. The transitions were chosen to avoid negative pressure, and thus cavitation damage. Head loss along the chute, air entrainment and freeboard determined the height of the walls. Figure 1.5 shows a view of the Itaipu spillway.

The specific design discharge of the chute is over $180 \, m^2 s^{-1}$, and the exit velocity is $40 \, ms^{-1}$. Such large values in both discharge and velocity dictate a trajectory spillway. The way of energy dissipation had to be selected carefully. Additional criteria involved the maximum recorded river level, the impact location of the jets at the centre of the river, and avoidance of hydraulic jumps in the bucket of the ski jump for low and medium discharges. Dispersion of the jets could be assisted with dispersion blocks, but these were not adopted due to the cavitation risk, difficulty of repair and complexity of construction. Accordingly, simple blocks were provided and a portion of hydraulic

Figure 1.5
Itaipu spillway
(Courtesy of N. Pinto)

energy is dissipated in the air prior to impact on the water cushion. An additional study of the Itaipu spillway was also conducted by Tarricone, et al. (1979).

After 10 years of operation Torales, et al. (1994) reported only minor *deterioration*. The maximum discharge on the spillway amounted to $40\,000\,\mathrm{m^3s^{-1}}$. Due to the continuous operation of the spillway during some years, some erosion was caused on the slopes downstream of the concrete slab which protects the flip bucket. Also, some significant scour of 8 to 10 m depth and over 10 m length occurred on the right slope 50 m downstream of the spillway.

Before initial operation of the spillway, the chutes and ski jumps were carefully inspected for finishing defects. Epoxy-mortar materials were used to patch up these defects. Some of these repairs were eroded away and epoxy-mortar was again used. At locations of the chute where rough mortar surfaces were observed, an epoxy-cement mix was applied to smooth out the rough surfaces. Intensive cavitation damage on the right wall of the left chute necessitated a modified deflector on the wall. Since 1985 the left chute performs properly.

Additional model studies were conducted to determine the energy dissipation of the chutes and the erosion in the Parana riverbed. This was necessary because of:

- high frequency of spillway operation,

- extreme concentration of discharge when less than three chutes in use,

- high unit discharge and velocity of the jets, and

- uncertainty of effect of tailwater cushion.

The *La Grande* hydroelectric complex is located on the La Grande river in Quebec (Canada). Its length is 850 km and the river discharges into James Bay. The total catchment area is almost $100\,000\,\mathrm{km^2}$.

Phase I of the La Grande complex was completed in 1985. Two reservoirs were created and their waters diverted into the La Grande river. Also, La Grande powerplants 2, 3 and 4 (from east to west) were constructed. Phase II started in 1987 and involves four additional plants: Brisay, Laforge 1, La Grande 2A, and La Grande 1. The main goal of phase II is to optimize operation of the diversions and reservoirs already created. Most of the reservoirs and impoundments required for the complex as a whole were created in phase I. A total of almost $18\,000\,\mathrm{MW}$ will be available in the year 2005. The hostile climate has an average annual temperature of $-4\,^\circ\mathrm{C}$ with wide variations between summer and winter. Floods occur twice a year, mainly in spring caused by snowmelt and also in autumn because of rainfall. The La Grande river falls 376 m along its course and the average annual discharge is $1700\,\mathrm{m^3s^{-1}}$. The rock in the vicinity of the plants is hard granite. Glaciers have removed surface rocks but much suitable construction material remains on the surface.

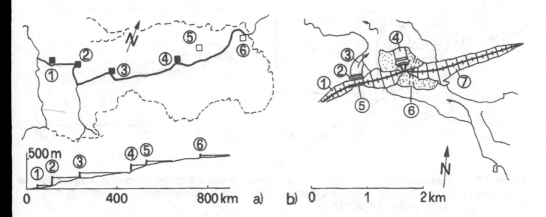

Figure 1.6 Plan of (a) James Bay complex with ① LG1, ② LG2, ③ LG3, ④ LG4, ⑤ Laforge 1, ⑥ Brisay, (---) reservoir boundary, (b) La Grande 3 dam with ① north dam section, ② powerhouse, ③ tailrace channel, ④ spillway, ⑤ intake structure, ⑥ diversion tunnels, ⑦ south dam section (Anonymous, 1990)

Figure 1.6(b) shows a plan of La Grande 3 (LG3). The site is characterized by a rock island dividing the river into two branches. The intake structure and the powerplant are located on the left branch, the spillway is on the top of the island and the diversion tunnels are on the right branch. The spillway is integrated in the rockfill dam extending on either side over a length of 3.8 km. The height of the dam is 93 m, its total volume 22×10^6 m^3 and additional 67 smaller dams with a total crest length of 23 km had to be constructed.

The spillway has a total capacity of 10 000 m^{3}s^{-1} and consists of five gated overflow structures each 12.2 m wide and equipped with a roller gate 20.6 m high. Two central piers and two lateral division walls are provided up to the end of the spillway to reduce separation of flow from the expanding spillway sidewalls. The water is discharged with a ski jump into the tailwater. Due to the slightly contracting spillway plan, from 78 m at the crest to 70 m at the centre and to 100 m at the ski jump, the lateral discharge distribution is uniform for almost all configurations of gate positions. The waters plunge into a pool excavated into the rock on the downstream side of the island. The difference of elevations between the maximum reservoir level (257 m a.s.l.) and the pool invert (168 m a.s.l.) is 89 m. The diversion channel from the plunge pool to the river is 213 m wide. Figure 1.7 shows plan and section of the spillway.

For optimum energy dissipation several features are included in the final spillway design:

- increase of spillway length to inhibit erosions close to dam,

- reduction of take-off elevation of ski jump to increase jet velocities, and

- shift of lateral guiding walls more to the sidewalls to improve distribution of jet discharge.

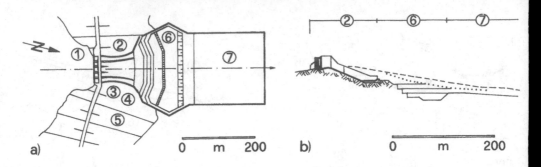

Figure 1.7 LG3 spillway (a) plan, (b) section. ① Approach walls, ② chute, ③ training wall, ④ side wall, ⑤ north dam section, ⑥ plunge pool, ⑦ spillway channel; natural ground along (- - -) spillway axis, (· · ·) west side boundary (Aubin, et al. 1979)

The final design was tested for discharges between 850 and 10 600 m^3s^{-1} and satisfactory flow behaviour both on the spillway and in the plunge pool was observed (Figure 1.8).

1.3   DESIGN DISCHARGE OF BOTTOM OUTLET

1.3.1   Purpose of bottom outlet

In many countries, the design of a bottom outlet is obligatory, and its inclusion is strongly recommended in all other instances. Primarily, a bottom outlet is a *safety structure*, and it can be used secondarily for the flushing of *sediment deposits* or for discharging surplus water, as outlined in Chapters 6 and 8. The load test of a dam occurs during the

Figure 1.8
LG3 spillway
(a) without
discharge (*Chantiers*
1991, **22**(10): 26), (b)
with 10000 m^3s^{-1}
during winter
(*Electric Power in
Canada* 1986)

a)

b)

first filling of the reservoir. Accordingly, this test has to be conducted with extreme care, because a failure has serious consequences (Chapter 10). The filling must be made progressively by accounting for the stability and the watertightness of both the dam, and its surroundings. A dam is thus typically filled to the first third of its height, and observations are conducted during several days or weeks. If the results are satisfactory, the dam is filled further, typically to two thirds of its height. Particular attention will be directed to the response of the dam under complete initial filling.

This *filling procedure* is only amenable if the reservoir level can be controlled, i.e. if a bottom outlet is available. The bottom outlet must thus be designed that the reservoir level can be kept constantly under arbitrary levels. Figure 1.9 shows the general configuration with a reservoir level at $(1/3)z$, where z is the maximum water depth of the reservoir. If Q_o is the reservoir approach flow and Q_b the design discharge of the bottom outlet, then $Q_b \geq Q_o$ during the test period. The discharge Q_o corresponds to an average value over a certain time

Figure 1.9
Design of bottom outlet for controlled reservoir filling, with an initial filling of, say one third of the dam height

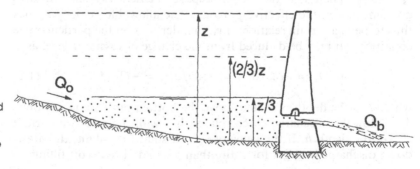

period, such as several weeks or even months. The choice of Q_o includes the storage effect of the reservoir, i.e. its surface and the tolerance, by which the level should be kept constant.

1.3.2 Drawdown of reservoir

The hydraulic pressure force on a dam increases with the second or third power of the water depth. A drawdown of the reservoir level is thus followed by a stress relief. If, for example, a full reservoir would be drawn down at $(2/3)z$, then the hydraulic pressure on the dam would be reduced by 46 to 76%. Accordingly, drawing down an endangered dam is highly efficient. The provision of water for power supply or drinking water may fail but the drawdown by the bottom outlet has to be fast. The hydraulic condition for the bottom outlet is thus $Q_b > Q_o$.

Obviously, a reservoir has to be emptied partly or entirely under normal usage, such as when revisions on the waterside of the dam or at the inlet of the bottom outlet are needed. Here, we concentrate on the *emergency drawdown* with three design guidelines:

1. rapid drawdown of a reservoir within the shortest time lapse, i.e. Q_b = maximum!,

2. the discharge Q_b may not damage the tailwater valley, i.e. $Q_b \leq Q_L$, where Q_L is the limit tailwater discharge, particularly during flood periods, and

3. the drawdown induces no shore slides, i.e. $Q_b \leq Q_s + Q_o$, where Q_s is the allowable discharge for shore protection.

The allowable discharge Q_s can be estimated with the limit drawdown velocity u_L and the reservoir surface A as $Q_s = u_L A$. The limit drawdown velocity is influenced mainly by geotechnic site conditions. If a reservoir is drawn down too rapidly, the seepage level along its shores cannot follow and the forces resulting in a slide are favoured. The flow of groundwater is mainly affected by the effective pore water velocity $v_s = ki/n_s$, where k is the Darcy seepage coefficient, i the hydraulic gradient and n_s the effective porosity. The allowable velocity u_L has thus to be in a certain relation to v_s, that depends on the particular site conditions. It may be deduced from the change of reservoir level as

$$u_L = -(\mathrm{d}z/\mathrm{d}t)_L = -(\mathrm{d}h/\mathrm{d}t)_L = -Q_s/A \qquad (1.1)$$

where h is the depth of water above the bottom outlet, and t is time. The allowable discharge Q_s correlates thus with the reservoir surface A, i.e. the depth h. If the velocity u_L is assumed constant, the drawdown discharge is larger for entire than for partial reservoir filling.

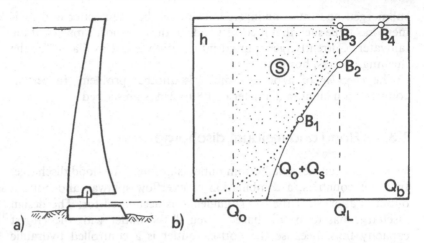

Figure 1.10
Optimization of
discharge-head
relation for design
discharge of
bottom outlet.
Domain of solutions
S and design points
B_1, B_2 and B_3

Figure 1.10 shows the design graph for the optimization problem. In the $h - Q$ diagram the restrictions are $Q_b \geq Q_o$, $Q_b \leq Q_L$, and $Q_b \leq Q_s + Q_o$, as previously established. The domain of maximum discharge is located on the right side of the domain of solutions S.

For flood conditions, the approach discharge Q_o is large and the limit discharge small. Then, the domain of solution shrinks, and can even vanish. During flood periods, a drawdown is thus impossible. For low discharge conditions, the domain of solution is large, however, and more options are possible. Usually, one would chose average discharge conditions for design, on which Figure 1.10 is based. It is stipulated that the cross-sectional area of the bottom outlet is given by design point B_1, and the corresponding discharge-head curve according to Torricelli is tangent to the domain of solution at point B_1. By incidence, the dotted curve on Figure 1.10 nearly touches also the design point B_3.

The *drawdown time* is a significant design element. Under emergency, one would like to draw down the reservoir level of say, the top third of the reservoir height, typically during a few days or weeks. The design of the bottom outlet should be checked also in this regard. The drawdown time can be integrated directly as a design condition for a suitable domain of solution. For reservoirs with a large storage volume, the drawdown time is dictated normally by restrictions. For reservoirs where the storage volume amounts to a multiple of the annual approach volume, drawdown times of months or even years may result. Then, the concept of rapid reservoir drawdown is of course invalid.

1.3.3 Flushing of sediments

In certain cases, bottom outlets are used for sediment flushing (Chapter 8). For the design of the outlet structure, the *continuity* of the sediment transport is important, because the material should reach the tailwater. If it deposits in the tailwater the bottom outlet may be submerged and thus endangered. The ratio between water and sediment has to be such that it corresponds to the transport capacity of the tailwater. Currently, generalized information is not available for the flushing process, however.

The *clogging* of a bottom outlet is another problem. In certain countries a minimum diameter of 1.8 to 2 m is prescribed.

1.3.4 Flood and residual discharge

In several countries, the bottom outlet is not used for flood discharge. In other countries, a combination of overflow spillway and bottom outlet is allowed, if the (N–1)-guideline is accounted for: The design discharge has to be discharged even if the outlet with the largest capacity fails. Because the bottom outlet is a controlled hydraulic structure, it can be added in the (N–1)-guideline. There exist even multiple bottom outlets with a capacity of the design discharge, and other overflow structures are not needed.

A bottom outlet designed according to previous criteria is normally too large to fit for the control of residual discharge. It had to be operated permanently with an extremely small opening not suited hydraulically. Therefore, a small outlet is often added to satisfy the needs for residual discharge. It resembles a conventional bottom outlet with an inlet, a tunnel, a gate and an outlet structure. This particular outlet can be designed as a mini-hydro plant, such that problems with the energy dissipation are solved.

1.3.5 Possibilities of failure

A bottom outlet may fail due to four causes:

- it cannot be opened,
- it opens uncontrolled,
- it clogs after opening, and
- it cannot be closed again.

The first two possibilities can nearly be excluded for modern designs of gates and their control. Some details of construction are described in Chapter 6. The failure due to the third possibility was addressed already in 1.3.3 by specifying a minimum diameter. Further details are given in Chapters 6 and 7. Clogging is mainly caused by mixtures of fine and coarse particles. The latter include wood, sediment and ice. Rock particles are particularly dangerous because they may sit in front of the outlet structure. The last possibility of failure can also be caused by coarse particles. A rock portion may be squeezed by the gate during

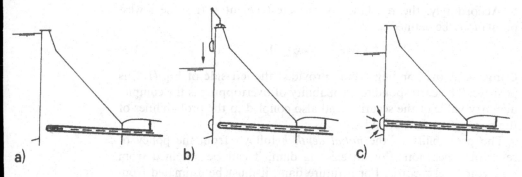

Figure 1.11   Example of procedures for building an emergency bottom outlet

closure and eventually moves only after several opening manoeuvres into the tailwater. Because of the absence of gate slots, sector gates are less prone to this phenomenon.

Given that the bottom outlet is the outlet structure with the lowest elevation, its failure causes serious problems of *revision* besides operational inconveniences. If a bottom outlet is clogged or if it cannot be opened access from the waterside is impossible. Divers are also in danger if the clogging suddenly dissolves. If the bottom outlet cannot be closed, it is inaccessible from the tailwater. From experience, an installation of an emergency gate has proved to be costly, when set up in such an emergency case. Figure 1.11 explains a procedure when the bottom outlet cannot be opened and the reservoir should be drawn down. A secondary tunnel is constructed from the tailwater, and a provisional gate is clapped at the position of breakthrough, which is installed from the dam crest. It is thus recommended to install an *emergency gate* in front of the guide gate right during the construction of the dam (Chapter 6).

1.4   DESIGN DISCHARGE OF SPILLWAY

1.4.1   Concept of crest height

A spillway is a safety structure against overflow. It should inhibit the overflow of water at locations which were not considered. The spillway is the main element for *overflow safety* and especially the safety against dam overtopping.

A structure is known to be safe against damage if the load is smaller than the resistance. What are the determining loads, and the resulting resistance with respect to the overflow safety? Are these crest heights, discharges or water volumes?

In the concept of crest height the load is composed of the sum of initial depth h, plus a depth increase r due to a flood, plus a wave depth *b*. Accordingly, the resistance of the crest height k (i.e. the lowest point) may be defined as

$$h + r + b - k \leq 0. \tag{1.2}$$

Conversely, overtopping occurs provided the left side of Eq. (1.2) is positive. The corresponding probability of overtopping is the complementary value of the security, and also coupled to the probabilities of the parameters h, r, b, and k.

The probability of the *initial depth h* follows from the policy of reservoir operation. For an existing dam, it can be obtained from reservoir level records. For a future dam, it must be estimated from the planned policy.

The probability of the *flood depth increase r* can be determined from flood routing. The result depends on the probabilities of the approach flood and the reservoir outflow. The probability of the approach flood depends on general flood parameters, such as maximum discharge, time to peak, time of flood, and flood volume. Regarding the reservoir outflow, the degree of aperture of the outflow structures has to be accounted for.

The probability of the *wave depth b* depends mainly on the characteristics of wind, its direction and the run-up slope of the dam relative to the water surface. In particular cases, ship waves and surge waves due to shore instabilities (including rock and snow avalanches, and land slides) are also included.

Normally, the parameter k is considered as a fixed number and not as a stochastic value, which is an acceptable assumption for concrete dams. For high earth dams, the probability of settlement could be introduced, but this is often omitted and a maximum settlement is accounted for.

The probability of a combination of the parameters h, r, b, and k thus mainly results in a variation of six basic variables: initial flow depth, peak flood discharge, time to peak, aperture degree, velocity and direction of wind. Assuming that these basic variables are *stochastically independent*, the result is straightforward. As an example, one could apply the Monte Carlo method to obtain a representative number of combinations, and thus a number s of $(h + r + b - k)$. If those sums with s equal or smaller than zero are labelled n, the safety q against overflow is

$$q = \frac{n}{s}. \tag{1.3}$$

If $m = s - n$ is the number of combinations where the sums are positive, then the *probability of overflow* is

$$p = \frac{m}{s}. \tag{1.4}$$

In the inset to Figure 1.12, those relations with a probability distribution are plotted for a fixed value of k.

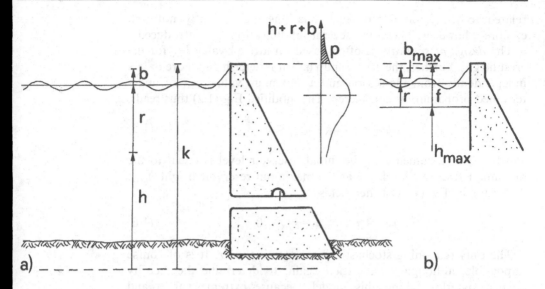

Figure 1.12 Dam with reservoir (a) initial depth h before flood flow, depth r of flood level, and wave height b, height of crest k, and probability of overtopping p, (b) maximum reservoir depth h_{max}, maximum run-up height b_{max} and freeboard f

The concept described is obvious, as it accounts for the relevant parameters: depth of reservoir and depth of crest. It can be used for sensitivity analyses and answer questions such as:

- What is the change of overtopping safety for changes in the reservoir operation? (relating to parameter h)

- What is the effect of time to peak for a given peak discharge? (relating to parameter r)

- What is the sensitivity of certain closing structures on the probability of failure? (relating to parameter r also)

- What is the result if the overtopping of a concrete dam by wind waves is accepted? (relating to parameter b)

- What is the effect of a crest height reduction due to damage, for instance? (relating to parameter k).

A set of stochastically independent parameters can normally not be assumed. In certain regions, large floods are combined with thunderstorms and flood waves are thus related to wind waves. In other regions, the floods have their origins far upstream from the reservoir and the accompanying winds may hardly reach the dam. Normally, dams are erected to store water during the rain period for the dry season. Accordingly, the reservoirs have usually reached a high level at the end of the rain period. The determining floods occur often at the end of the rain period, such that the initial depth is stochastically related to the flood depth. Such relations are currently not well explored, however. Therefore, several simplifications are introduced.

The *design of spillways* is often based on a fixed value k_{fix} for the crest height. Further, the wave runup height is determined more or less independently from the season and a maximum value b_{max} of undetermined probability is considered. The modified Eq. (1.2) thus reads

$$h + r + b_{max} - k_{fix} \leq 0. \tag{1.5}$$

Further, it is assumed that the initial reservoir level is equal to the maximum reservoir level, i.e. at the maximum reservoir height h_{max}. Accordingly, Eq. (1.5) further yields

$$h_{max} + r + b_{max} - k_{fix} \leq 0. \tag{1.6}$$

The only remaining stochastic parameter is thus r. It is of course impossible, although always tried again, to determine the security against overflow along this model, because extreme values and stochastic values cannot simply be superposed. An exception is the reservoir where surface waves can be neglected and where practically no surface fluctuations occur, such that $h = h_{max}$.

Another popular approach uses instead of the parameters k_{fix} and h_{max} the *freeboard height*

$$f = k_{fix} - h_{max} \tag{1.7}$$

and requires that

$$r \leq f - b_{max}. \tag{1.8}$$

1.4.2 Concept of water volumes

The heights h, r, b, and k are related to particular volumes of a reservoir, such as

V_h = reservoir volume at initial reservoir level,

V_r = flood volume minus outflow volume, corresponding to the flood storage volume,

V_b = wave run-up volume,

V_k = maximum reservoir volume up to dam crest level.

Eq. (1.2) may be written analogously with regard to the volume V as

$$V_h + V_r + V_b - V_k \leq 0, \tag{1.9}$$

and all concepts presented earlier can simultaneously be transposed.

An alternative approach results if the variables depending on a particular event are compared. Both V_k and V_h are independent, and V_b too can often be regarded as an independent variable. If these three

parameters are substituted as the available storage volume
$V_R = V_k - V_h - V_b$, then, from Eq. (1.9)

$$V_r < V_R. \tag{1.10}$$

Therefore, the hydrologic load of the reservoir in terms of the *needed*
storage volume V_r can be opposed to the *available* storage volume V_R.
Both, V_r and V_R are stochastical variables. Accordingly, the safety
against overflow, and the overflow probability can be derived from the
probabilities in V_r and V_R, as described previously.

If the following assumptions are introduced, as also presented earlier:

$V_h = V_{h_{max}}$ reservoir volume under full reservoir,

$V_{b_{max}}$ = reservoir volume under maximum wave run-up,

$V_{fix} = V_{k_{fix}} - V_{h_{max}}$ = volume corresponding to freeboard height f,

then in analogy to Eq. (1.8)

$$V_r \leq V_f - V_{b_{max}}. \tag{1.11}$$

Again, the security against overtopping cannot be predicted, if singular
components such as $V_{h_{max}}$ are introduced as maximum values. The
exception are dams, which are permanently full, and where surface
waves are insignificant.

1.4.3 Concept of discharges

The effectively needed storage volume V_r can be determined from the
mass balance as

$$V_r = \int_T (Q_a - Q_z)\mathrm{d}t = \int_T Q_r\mathrm{d}t \tag{1.12}$$

where Q_a = reservoir inflow during the filling time T,
             Q_z = reservoir outflow during the same time T,
             Q_r = reservoir storage.
The relation with the reservoir heights h is

$$Q_r = A\frac{\mathrm{d}h}{\mathrm{d}t} \tag{1.13}$$

where $A = A(h)$ is the reservoir surface. The filling time ends when
$Q_z = Q_a$, i.e. $Q_r = 0$. Eq. (1.10) can therefore be written as

$$\int_T Q_z\mathrm{d}t - \int_T Q_a\mathrm{d}t \leq V_R \tag{1.14}$$

or

$$V_z - V_a \leq V_R \tag{1.15}$$

where V_z = reservoir inflow volume during time T,
$\quad\;\; V_a$ = corresponding reservoir outflow volume.

In Figure 1.13, these relations are explained. The probabilities of $(V_z - V_a)$, and V_R may be used to determine the securities of overflow and the relating probabilities. These computations become again useless if fixed values instead of stochastic values are admitted. A popular example involves the so-called $(N-1)$-condition for the reservoir outflow. Instead of relating the availability of a regulated outflow to probability, one is typically faced with a situation such as: out of the N outlets there is *one* (and the one with the largest capacity) not

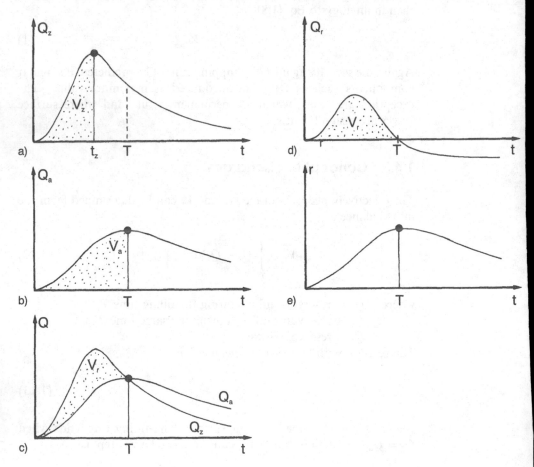

Figure 1.13 Hydrographs of (a) inflow flood $Q_z(t)$, (b) reservoir outflow $Q_a(t)$, (c) superposition and required storage volume V_r until end of time of rise T, (d) reservoir storage $Q_r(t)$ and (e) increase of storage $r(t)$. (•) maximum of functions

available. Other examples, such as fixing the maximum reservoir depth h_{max} as the initial depth for floods have already been mentioned.

The most important parameter is the *flood wave*, such as shown in Figure 1.13, and characterized with the peak discharge $Q_{z_{max}}$ and time to peak t_z. The temporal wave profile is then given as an empirical function $Q_z(t_z)$. The determining flood for the reservoir volume and the spillway structure is called *design flood*.

1.4.4 Design assumption

Numerous dam failures due to overtopping point to the significance of the design flood. Its resulting probability of failure has to be minimum, at least infinitely small. Because the lifespan of a usual dam is of the order of one hundred years, the probability of failure has to be related also to this period. If a value of 1%, or 0.1% for the entire lifespan is assumed, then the probability of failure *per year* is 10^{-4}, and 10^{-5}, respectively. For a large potential of damage, one would choose one or two orders of magnitude smaller. How can such small values of probability be guaranteed? It is not the purpose here to outline the corresponding philosophies of each country worldwide.

Computational or Intuitive Approach

In section 1.3.1, the determination of safety against dam overtopping and the probability of failure were described. The *probability of failure* can be computed by combining the parameters mentioned with their probability of occurrence, such that the probability of overtopping at the lowest point of the dam crest can be determined. The difficulty of the procedure is in the estimation of the probabilities of some parameters. For dams, which have existed for several decades and whose safety against overflow is permanently checked, the wave runup has eventually been observed and the availability of the outflow structures is known. Also, information regarding the height of initial flow depth before floods is probably available. For new dams, estimations have to be advanced, however. Mention was already made that some parameters under consideration are stochastically dependent, which adds further complications.

The probability of failure cannot be determined if some stochastic parameters are assumed to have a certain probability, and others are considered as fix values. The latter correspond normally to intuitively chosen maxima. Such a mixed approach is currently the common approach, however, and there is nothing to counter as long as no probability computation is performed.

Normally, the *mixed approach* involves a rare design flood, for which considerations of probability are still appropriate. As an example, a 1000-year flood is chosen. Then, intuitive security factors on parameters such as the initial reservoir outlets are introduced. As mentioned earlier, the maximum reservoir elevation is often set equal to a maximum reservoir level, and the (N–1)-condition is added in relation to the reservoir outlets. Further, an immobile power plant is considered, i.e. the related power plant, the water supply station, or irrigation works are switched off, and the freeboard is chosen higher than the maximum wave run-up. With these additional safety measures, a probability of overtopping well below 0.1% within 100 years may be achieved, i.e. a value of practically zero.

Design Flood

The design flood is a reservoir inflow of extremely small probability, of 1000 or even 10 000 years of occurence, as previously mentioned. To estimate these rare values, a data series is evidently not available. There are conventions, however, by which *extrapolations* can be made based on a data series of several decades. These extrapolations of, say, the reservoir inflow discharge or the rainfalls are difficult to interpret. They include knowledge of local particularities, and a detailed hydrologic approach. The times when some flood discharge formulae have been applied without particular reference to a catchment area have definitely passed. As an engineer would hardly transpose the geology of one dam site to the other, it is impossible to use hydrologic data to cases other than considered.

Actually, two different design cases are used in many countries, considering a *smaller* and a *larger* design flood. The *smaller design flood* has a return period of the order of 1000 years. It must be received and diverted by the reservoir without damage. Often, a full reservoir level is assumed and all intakes for power plants etc. are blocked, and (N–1) spillway outlets are in operation. Whether the bottom outlet can be accounted for diversion is a question, but there is a tendency to include it in the approach. The freeboard is specified and must be observed.

For the *larger design flood* a return period of 10 000 years is considered, for example. As such an extraordinary event can be extrapolated from the limited data available only as a rough estimation, other conventions are used. One approach increases the 1000 year flood by 50% both in peak discharge and time to peak. Another approach is based on the concept of the *possible maximum flood* (PMF). Accordingly, a rainfall-runoff model with the most extreme combination of basic parameters is chosen, and no return period is specified. This design flood has to be diverted without a dam breaching. However, small damages at the dam and the surroundings may occur. Therefore, the conditions for wave run-ups and the availability of N–1 outlets, among others, are not entirely satisfied.

In some countries, the definition of the design flood is also related to the *potential damage* due to a dambreak. As the prediction of such a potential for the next century is difficult, hardly any figures are given. A usual compromise is to increase the design quantity if high dams and large reservoir volumes are involved.

Design Flood of Spillway Structure

The design flood $Q_{a_{max}}$ of the spillway may be determined from Eq. (1.14), that is from the inflow design flood, and includes the described effects of initial reservoir depth, reservoir freeboard, and availability of outflow structures.

1.5   DESIGN DISCHARGE OF INTAKE STRUCTURE

1.5.1   Purpose

Dams have essentially the following four main purposes:

- retain water to increase the hydraulic pressure and use it for power generation or diversion to other pressure systems,
- creating waters with a quiet surface for navigation, recreation or environmental use,
- flood retention, and
- store water to balance natural afflux and human demands of drinking water, irrigation and water power.

The following aims at discussing the design basis for the latter purpose in particular.

1.5.2   Storage characteristics

Figure 1.14 shows the relation between the reservoir elevation z and the reservoir surface A, and the reservoir volume J, respectively, based

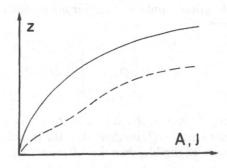

Figure 1.14
Relation between (- -) reservoir surface A, and (—) reservoir volume J as functions of reservoir filling z

on the reservoir topography. The function $A(z)$ can be determined from a topographical map, and the function $J(z)$ follows from the integration $J = \int A(z)\mathrm{d}z$. It can often be approximated with a power function $J = az^b$ where b is a characteristic of the reservoir shape. Usually, the domain of b is between 1.3 and 4.4 (Vischer and Hager, 1992). Reservoirs with a nearly rectangular section have smaller values of b, and those with a nearly triangular shape have larger values of b.

1.5.3 Reservoir storage

The hydraulic characteristics of a reservoir are described by the storage equation

$$\frac{\mathrm{d}J}{\mathrm{d}t} = A\frac{\mathrm{d}z}{\mathrm{d}t} = Q_o - Q_b \tag{1.16}$$

where Q_o is the reservoir inflow and Q_b the reservoir outflow, with t as time. The outflow includes also storage losses due to transpiration and seepage. During a period T, Eq. (1.16) thus reads

$$J_o + \int_0^T Q_o\mathrm{d}t - \int_0^T Q_b\mathrm{d}t = J_T. \tag{1.17}$$

Typically, the period T corresponds to an average hydrologic year, where $J_o = J_T$ and thus

$$\int_0^T Q_o\mathrm{d}t = \int_0^T Q_b\mathrm{d}t. \tag{1.18}$$

The reservoir storage is thus assumed equal at the beginning and at the end of a year, with corresponding reservoir in- and outflows.

In general, the reservoir inflow is given by nature and cannot be modified. The reservoir outflow in turn can be controlled with an *outflow policy*. Obviously, the reservoir level is not allowed to fall and rise beyond the minimum (subscript m) and maximum (subscript M) reservoir elevations, and the reservoir management is based on the following restrictions

$$J_m < J_o + \int_0^t Q_o\mathrm{d}t - \int_0^t Q_b\mathrm{d}t < J_M \tag{1.19}$$

where, evidently, $J_m < J_o < J_M$ and $J_m < J_T < J_M$.

If the reservoir outflow Q_b is released by the intake structure and the reservoir losses are negligible, then the difference $J_M - J_m$ is referred to the useful volume ΔJ. If, however, the outflow is partly released by the spillway structure, then the useful outflow is only a portion of ΔJ, with the remainder as a portion of excess volume. Accordingly, each reservoir user aims at reducing the spillway discharge to the minimum. This is possible only for large reservoirs with a large intake structure.

Figure 1.15 describes the typical storage procedure: The hydrologic year can often be divided into a first period with a large and a second period with a small reservoir inflow. At the beginning of the hydrologic year, the reservoir has a minimum elevation and is filled subsequently. Accordingly, much water is stored in the reservoir during the first period. It can then be used in the second period for purposes such as irrigation, hydropower and water supply. If in the first period the inflow exceeds the storage capacity, the spillway comes into operation. Figure 1.15 refers to an average hydrologic year, during which an overflow may occur. The design of a reservoir thus involves a complicated *optimization procedure* that is not detailed here.

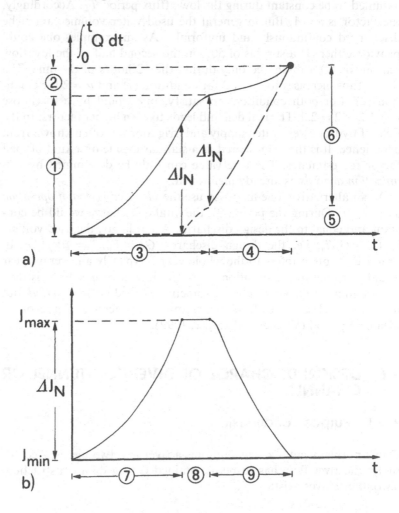

Figure 1.15
Balance of reservoir
(—) in- and (---)
outflows during
average year
provided the
storage water is only
used during the low
flow period.
(a) Summation
curve of reservoir in-
and outflow, with
inflow during
① storm period, ②
dry period, ③ rain
period, ④ dry
period, ⑤ excess
discharge, ⑥ water
for use,
(b) hydrograph of
reservoir volume
with ⑦ time of filling,
⑧ time of full
reservoir, ⑨ time of
emptying

1.5.4 Design principle and standard values

It is usual and useful to design a power plant, a water supply station or irrigation works from the inlet to the outlet structures based on the same design discharge Q_D. This discharge is, therefore, practically equal to the discharge capacity of such an hydraulic facility, i.e. $Q_b \leq Q_D$. Provided the entire outflow of the reservoir is used by the facility, i.e. if the spillway and the bottom outlet are not in operation, the design discharge is logically $Q_D = Q_{b_M}$. The design discharge can be determined from

$$Q_D = \nu \frac{V_u}{T_L}. \qquad (1.20)$$

V_u is the used water volume, T_L the duration of the low afflux period and ν is a factor. In Figure 1.15 for instance the intake discharge is assumed to be constant during the low afflux period T_L. Accordingly, the factor is $\nu = 1$. But in general the used water volume cannot be discharged continuously and uniformly. As an example, one could provide either (1) a surplus of 50% in the second half of the low flow season, or (2) a discharge only during the 12 hours of daytime. The factor thus increases to $\nu = 1.2$ for condition (1) and $\nu = 2$ for condition (2). For both conditions to satisfy, one would have to choose $\nu = 1.2 \times 2 = 2.4$. The real demand leads to even higher factors. In the field of hydropower, water supply and irrigation ν is often known from experience. It is then considered as a standard value or a kind of load factor for estimates. The true value can only be determined by *optimization analysis* as already mentioned.

As an alternative, one may also use the *idealized period of operation* $T_i = T_L/\nu$. During the period T_i, the intake discharge would be constant and equal to the design discharge. Accordingly, the used volume is $V_u = Q_D T_i$, i.e. the design discharge $Q_D = V_u/T_i = \nu V_u/T_L$. In practice, T_i often refers to the whole year – typically an average year – and not only to the duration T_L of the low flow season. T_i is then expressed in hours per year and is sometimes called *full load period*. But it is nevertheless an artificial figure that is a reference figure or an idealized period (Vischer and Hager, 1992).

1.6 DESIGN DISCHARGE OF DIVERSION TUNNEL OR CHANNEL

1.6.1 Purpose of diversion

Dams normally cut a valley, and thus a river into two pieces. Accordingly, the river flow has to be maintained during dam construction. Two alternatives exist:

Figure 1.16
① Diversion tunnel
during the
construction period
of a dam with
cofferdams,
② upstream and
③ downstream of
the dam. ④ dam
axis and ⑤ plug
after dam
construction

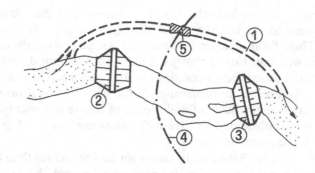

- temporary river diversion by one or several tunnels, (Figure 1.16), or
- dam construction by stages, to ensure an artificial breach.

A combination of both measures is also possible.

Because a diversion has to be designed for flood discharge, it corresponds practically to a spillway, that must divert the *construction flood*. But in addition to a conventional overflow structure it has to divert not only water but also the sediment load.

1.6.2   Concepts of design

In practice, three concepts of design for the diversion structure can be distinguished:

1. The entire construction flood is diverted by the diversion tunnels,

2. the entire construction flood is diverted by the artificial dam breach, or

3. the construction flood is diverted up to a limit value by the diversion tunnels, and by the artificial dam breach for larger discharges.

Concept 1 is favoured naturally, because the construction progress of the dam is not disturbed. However, the design discharge has to be estimated quite correctly, and the tunnels have to be not too costly due to their diameters and lengths. Accordingly, concept 1 suits for relatively small design discharges and tunnels with a sufficient bottom slope. As an example, most dams in the Swiss Alps have been erected with diversion tunnels.

Concept 2 demands a permanent opening of the artificial dam breach and disturbs, to a certain degree, the construction progress of the dam. It is applied where a diversion tunnel is uneconomical, such as for large design discharges and small bottom slopes. The artificial dam breach can be considered as a short channel across a construction area designed for the construction flood. Accordingly, the latter must be predicted quite accurately again. Yet, a channel has in principle a larger capacity than a tunnel.

Concept 3 is applied where both a tunnel or a dam breach are uneconomic alone because the hydrologic situation is difficult to assess. Then, for discharges below a limit discharge only the diversion tunnel is working. For discharges in excess of the limit discharge, the dam breach comes into operation, or the construction area is submerged. As a result, a disturbance of the constructional progress or damages can occur. Such a flood should of course not lead to serious floods in the tailwater. Thus, the cofferdams as shown in Figure 1.16 have to be designed for overflow.

Clearly, concept 3 is easier to realize for concrete dams than for rock or earth dams. The latter in particular cannot be made safe against overflow in arbitrary phases of construction. As an example, the design discharge during construction of the concrete dam of Cabora-Bassa on the Zambesi river (Mozambique) was estimated at $12\,000\,\mathrm{m^3s^{-1}}$, and the capacity of the diversion tunnel was fixed at only $4500\,\mathrm{m^3s^{-1}}$. Consequently, an overflow risk of the construction site was accepted, and various measures were taken to counter damages, such as the floodproofing of the cofferdams previously mentioned.

1.6.3 Choice of design flood

As for the choice of the design flood for the overflow structure (Figure 1.4), elevations of water level are relevant. Cofferdams, wingwalls and construction limits may, or may not, be flooded. The corresponding probability of flooding can be defined as for spillways. Flooding occurs provided that the sum of bottom elevation s, flow depth h and wave height w is larger than the elevation k of the shore line (Figure 1.17a). Note that the bottom elevation may vary with sediment transport and sand banks.

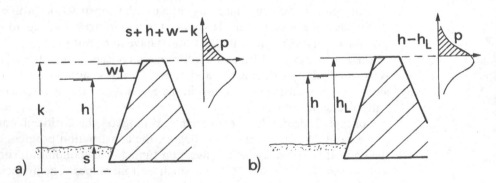

Figure 1.17 Schematic section of a cofferdam with the bottom elevation s, flow depth h, wave height w and crest height k. (a) Probability distribution of the function $(s + h + w - k)$ and flooding probability p; (b) probability distribution of the function $(h - h_L)$ and flooding probability p, with h_L as limit depth

The determination of the *flooding probability* is normally simplified by:

- considering both parameters s and k as stochastic constants, i.e. the effect of deposits or scour is excluded by constructional means. Also, settlements of the cofferdam are assumed known, and

- neglecting the effect of surface waves.

Consequently, flooding occurs provided the flow depth h is larger than the limit depth h_L for the design discharge (Figure 1.17b).

If the relation between the limit depth h_L and the corresponding limit discharge Q_L can be predicted, especially by approximating the roughness effect, the probability of flooding can be expressed with the probability of floods. If the limit discharge Q_L has a return period of n years, the probability of flooding within one year is

$$p_1 = n^{-1}, \tag{1.21}$$

and for m years of construction (for a diversion that does not change during construction)

$$p_m = 1 - (1 - n^{-1})^m. \tag{1.22}$$

This probability is also referred to as the *hydrologic risk* of the construction site. Table 1.1 gives some numbers for p_m as a function of the return period. For $n \gg m$, one would obviously have $p_m = m/n$. If one would like to limit the risk of flooding to say 10%, the design flood has a return period of $n = 10\,m$. The same return period could also be attributed to the construction flood discharge. Accordingly, the design discharge would amount to $Q_D = Q_L = Q_n$ where $n = 10\,m$. Whether 10% or only 1% is admitted follows from a *risk analysis* or at least from an estimation of the risk considering the economic consequences.

Table 1.1 Hydrologic risk for a construction period of 5 years, with n as return period of the limit discharge, and p_m as probability of flooding.

n	[year]	10	20	50	100	200
m/n	[%]	50	25	10	5	2.5
p_m	[%]	41	23	9.6	4.8	2.5

REFERENCES

Anonymous (1990). Intensive hydro construction resumes in Quebec. *Water Power & Dam Construction*, **42**(1): 28–35.

Aubin, L., Champoux, R., Sachter, L., Alam, S. (1979). L'évacuateur de crue de LG3. *13 ICOLD Congress* New Delhi **Q50**(R7): 105–119.

De Moraes, J., Rodriguez Villalba, J., De Mello, W.F., Poly, L.C., Berny, O. Acosta, G., Sarkaria, G.S. (1979). Selection of basic design of Itaipu spillway. *13 ICOLD Congress* New Delhi **Q50**(R14): 249–272.

Stutz, R.O., Giezendanner, W., Ruefenacht, H.P., (1979). The ski jump spillway of the Karakaya hydroelectric scheme. *13 ICOLD Congress* New Delhi **Q50**(R33): 559–576.

Tarricone, N.L., Neidert, S., Bejarano, C., Fonseca, C.L. (1979). Hydraulic model studies for Itaipu spillway. *13 ICOLD Congress* New Delhi **Q50**(R43): 749–766.

Torales, M.A., Villalon, O., Szpilman, A., Cardozo, F., Rosco, J.A., Piasentin, C., Fiorini, A., (1994). Itaipu spillway deterioration and maintainance after 10 years of operation. *18 ICOLD Congress Durban* **Q71**(R38): 573–588.

Vischer, D. and Hager, W.H. (1992). *Hochwasserrückhaltebecken* (Flood retention basins). Verlag der Fachvereine VdF: Zürich, Switzerland (in German).

Zanon, A. (1988). The Karakaya hydro-electric plant on the Euphrates river in Turkey. *Idrotecnica* **15**(2): 205–212.

Spillway of Grand Coulee concrete dam, USA (Water for California)

2

Overflow Structures

2.1 INTRODUCTION

2.1.1 Overflow gates

The overflow structure has a hydraulic behaviour that the discharge increases significantly with the head on the overflow crest. Nevertheless, the height of the overflow is usually a small portion of the dam height. Further, gates may be positioned on the crest for *overflow regulation*. During floods, and if the reservoir is full, the gates are completely open to promote the overflow. A large number of reservoirs with a relatively small design discharge are ungated (Figure 2.1).

Currently, most large dams are equipped with gates to allow for a flexible operation. The cost of the gates increases mainly with the magnitude of the flood, i.e. with the overflow area. Improper operation or malfunction of the gates is the major concern which may lead to serious overtopping of the dam. In order to inhibit floods in the tailwater, gates are to be moved according to gate regulation. Gates should be checked against *vibrations*. The advantages of gates at overflow structures are:

- variation of reservoir level,
- flood control,
- benefit from higher storage level

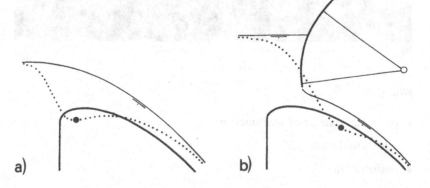

Figure 2.1
Overflows structure
(a) ungated and
(b) gated with (···)
bottom pressure
profile and
(o) minimum
bottom pressure

a)　　　　　　　　　　　　　b)

a)

b)

Figure 2.2
Photograph of
ungated and gated
spillway, (a) Shiroro
dam, Nigeria, and
(b) Rio Grande
dam, Colombia
(Courtesy Torno
S.p.A. Italy)

whereas:

- potential danger of malfunction,
- additional cost, and
- maintenance

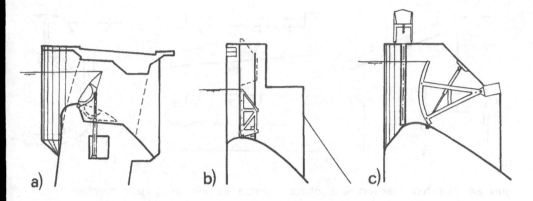

Figure 2.3   Gate types (a) flap gate, (b) vertical gate and (c) radial gate

are disadvantages to be considered. Depending on the size of the dam and its location, one would prefer gates for:

- large dams,
- large floods, and
- easy access for gate operation.

Three *types of gates* are currently favoured (Hartung, 1973; ICOLD, 1987): hinged flap gates, vertical lift gates and radial gates (Figure 2.3). The *flaps* are used for a small head of some metres and may span over a considerable length. The *vertical gate* can be very high but requires substantial slots, a heavy lifting device and an unappealing superstructure. Currently, those gates are equipped with fixed wheels, and may have flaps at the top. The *radial gates* are most frequently used for medium and large overflow structures because of their simple construction, the modest force required for operation and the absence of gate slots. They may be up to 20 m × 20 m, or also 12 m high and up to 40 m wide. The radial gate is limited by the strength of the trunnion bearings.

The risk of gate jamming in seismic sites is relatively small if setting the gate inside a stiff one-piece frame. For safety reasons, there should be a number of moderately sized gates rather than a few large gates. For the overflow design it is customary to assume that the largest gate is out of operation. The regulation is ensured by hoists or by hydraulic jacks driven by electric motors. Stand-by diesel-electric generators should be provided if power failures are likely (ICOLD, 1987).

2.1.2   Overflow types

Depending on the site conditions and the hydraulic particularities an overflow structure can be of various designs. Figure 2.4 shows a selection of overflow type structures, involving mainly the:

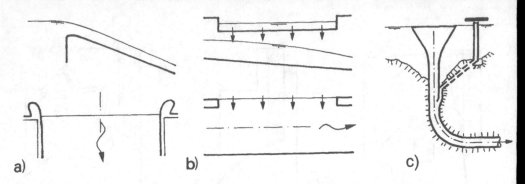

Figure 2.4 Main types of overflow structures (a) frontal, (b) side, and (c) shaft overflow

- frontal overflow,
- side channel overflow, and
- shaft overflow.

Other types of structures such as the *labyrinth spillway* use a frontal overflow but with a crest consisting of successive triangles or trapezoids in plan view. Still another type is the *orifice spillway* in the arch dam such as the Kariba dam in Zimbabwe (Figure 2.5). There exists, of course, the potential of orifice blockage by debris. Also, the design discharge must be known exactly because overtopping may occur. The orifice is governed by an insensitive discharge-head equation compared to the overflow structure. The orifice structure demands a high degree of reliability and control from the hydromechanical equipment. Also, the properties of material used must be exactly known under the effect of high velocity flow. Additional provision for the flood safety is thus necessary. The non-frontal overflow types are used for small and intermediate discharges, typically up to design floods of $1000 \, \mathrm{m^3 \, s^{-1}}$.

The *shaft type spillway* was developed in the 1930s and has proved to be especially economical, provided the diversion tunnel can be used as a

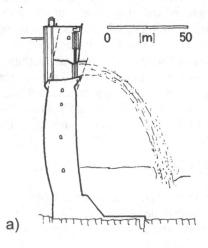

Figure 2.5
Orifice type dam
(a) section
(b) photograph of
Kariba dam (*La
Technique des
Travaux* 1962 **38**:37)

b)

Figure 2.5   *(Continued)*

tailrace. The structure consists of three main elements, namely the intake, the vertical shaft with a 90° bend, and the almost horizontal spillway tunnel. Air by aeration conduits is provided in order to prevent cavitation damage at the transition between shaft and tunnel. Also, to account for flood safety, only non-submerged flow is allowed such that free surface flow occurs along the entire structure from the intake to the dissipator. The hydraulic capacity of both the shaft and the tunnel is thus larger than that of the intake structure. The system intake-shaft is also referred to as 'Morning-glory overflow' due to the similarity with a flower having a cup shape. Figure 2.6 refers to the Hungry Horse spillway with a design discharge of $1500 \, \text{m}^3 \, \text{s}^{-1}$. The shaft has a 45° slope and is thus non-standard. Cavitation damage occurred at the transition from the sloping to the horizontal tailrace tunnel.

The *side channel overflow* was successfully used at the Hoover dam (USA) in the late 1930s. The arrangement is advantageous at locations where a frontal overflow is not feasible, such as for earth dams, or when a different location at the dam side yields a better and simpler connection to the stilling basin. Side channels consist of a frontal type overflow structure and a spillway with axis parallel to the overfall crest. The specific discharge of the overfall structure is normally limited to $10 \, \text{m}^3/\text{sm}$, but for lengths of over 100 m. A classic overflow type structure is located at the Hoover dam with the particularity that the flow is taken to the dissipator in a spillway tunnel. Many other side

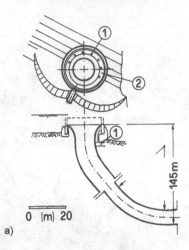

a)

b)

Figure 2.6
Shaft type spillway
(a) section and plan
with ① aeration
conduits,
② ring gate,
(b) photograph
(*Engineering News-Record* 1953
Nov.5: 46)

channels have been built worldwide, but this overflow type is restricted to small and medium discharges.

The *frontal type overflow* is the standard overflow structure, both due to its simplicity and the direct connection of reservoir to tailwater. It can normally be used in both arch and gravity dams. Also earth dams and frontal overflows can be combined, with particular attention against overtopping. The frontal overflow can easily be extended with *gates and piers* to regulate the reservoir level, and to improve the approach flow to the spillway. Gated overflows of 20 m gate height and more have been constructed, with a capacity of 200 m³ s⁻¹ per unit width. Such overflows are thus suited for medium and large dams, with large floods to be conveyed to the tailwater. Particular attention has to

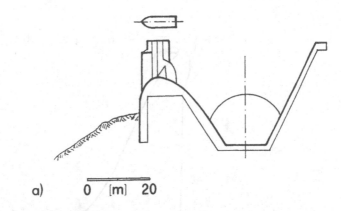

a)   0 [m] 20

Figure 2.7
Side channel type
overflow
(a) section,
(b) photograph of
Hoover dam, USA
(US Dep. of the
Interior, Bureau of
Reclamation, The
Reclamation
Engineering Center:
Denver, 1950)

b)

be paid to cavitation due to the immense heads that may generate pressure below the vapour pressure in the crest domain. Also, the gate piers have to be carefully shaped in order to obtain a symmetric approach flow. Figure 2.8 shows a typical frontal overflow.

The downstream portion of a frontal overfall may have various shapes. Usually, a *spillway* is connected to the overfall crest as a transition between overflow and energy dissipator (Figure 2.9a). Also, the crest may abruptly end in arch dams to induce a falling nappe that impinges on the tailwater (Figure 2.9b). Another design uses a cascade spillway to dissipate energy right away from the crest end to the tailwater, such that a reduced stilling basin is needed (Figure 2.9c). The standard design involves a smooth spillway that conveys flow with a high velocity either directly to the stilling basin, or to a trajectory bucket where it is ejected in the air to promote jet dispersion and reduce the impact action. Figure 2.10 refers to these two standard types of energy dissipation.

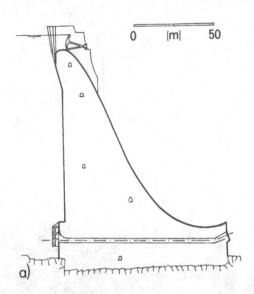

a)

b)

Figure 2.8 Frontal type overflow (a) section, (b) photograph of Aldeadavila dam, Spain (ICOLD **Q33**, R22)

Figure 2.9
Connection
between frontal
overflow and
dissipator (a)
chute spillway, (b)
free fall, (c)
cascade

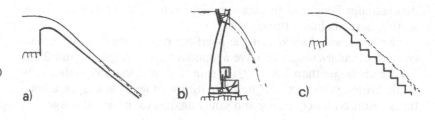

a)   b)   c)

Figure 2.10
Connections
between spillway
and tailwater
(a) stilling basin,
(b) trajectory basin

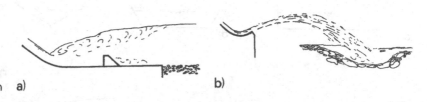

a)   b)

2.1.3   Significance of overflow structure

According to ICOLD (1987) the overflow structure and the design
discharge have a strong impact on the dam safety. Scale models of
overfall structures are currently needed in cases where (ICOLD, 1987):

- the valley is narrow and the approach velocity large such that an
  asymmetric flow pattern develops,
- the overflow and pier geometry is not of standard shape, and
- structures at either side of the overflow may disturb the spilling
  process.

For all other cases, the design of the overflow is so much standardized
that no model study is needed, except for details departing from the
recommendations.

2.2   FRONTAL OVERFLOW

2.2.1   Crest shapes

Overflow structures of different shapes are shown in Figure 2.11. In
plan view, the crest can be straight (standard), curved, polygonal or of
labyrinth shape. The latter structure has an increased overflow capacity with regard to the width of the structures.

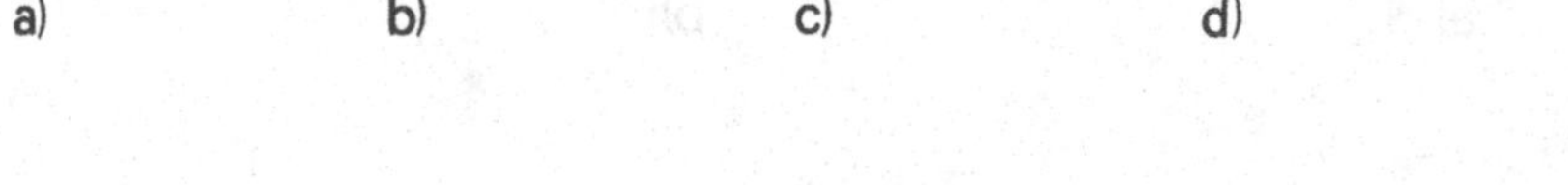

Figure 2.11
Plan views of
overflow structures
(a) straight,
(b) curved,
(c) polygonal,
(d) labyrinth

The *transverse section* of an overflow structure may be rectangular,
trapezoidal, or triangular (Figure 2.12). In order to have a symmetric

a)   b)   c)   d)

downstream flow, and to accommodate gates, the rectangular cross-
section is used almost throughout.

The *longitudinal section* of the overflow can be made either broad-
crested, circular-crested or have a standard crest shape (Figure 2.13).
For heads larger than 3 m, say, the standard overflow shape should be
used. Although its cost is higher than for the other crest shapes, advantages result both in capacity and safety against cavitation damage.

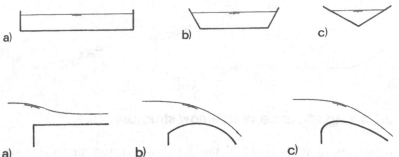

Figure 2.12
Transverse sections
of overflow structure
(a) rectangular,
(b) trapezoidal,
(c) triangular

Figure 2.13
Longitudinal section
of overflow structure
(a) broad-crested,
(b) circular-crested,
(c) standard-
shaped

2.2.2 Standard crest

The flow over a structure involves curved streamlines with origin of
curvature below the flow. The gravity component of a fluid element is
thus reduced by the centrifugal force. If curvature is sufficiently large the
internal pressure may drop below the atmospheric pressure and even
attain values below the vapour pressure for large structures. Then,
cavitation may occur with a potential *cavitation damage*. Given the
importance of the overflow structure, such conditions are unacceptable.

For medium and large overflow structures, the crest shape of a fully
aerated *thin-plate weir* is adopted, because the resulting overflow jet
corresponds to the 'natural' shape and involves atmospheric pressure
both along the lower and upper nappe boundaries. The basis of the
crest shape is well defined and involves a smooth thin plate weir of a
certain angle relative to the vertical. Of particular interest is the vertical
weir shown in Figure 2.14. The crest shape should be knife sharp, with
a 2 mm horizontal crest and a 45° downstream bevelling. In order to

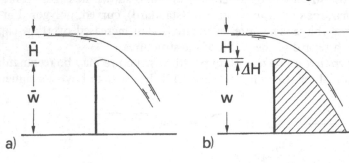

Figure 2.14
(a) Thin-plate weir
fully aerated,
(b) corresponding
standard crest

inhibit scale effects due to viscosity and surface tension, the head on the weir should be at least 100 mm, and the height of weir has to be at least twice as large as the maximum head. Then, effects of approach velocity are insignificant.

Figure 2.15 refers to thin-plate weir flow and the corresponding overflow structure at design flow. The lower nappe separates at the upstream crest edge from the weir plate, rises to a maximum trajectory point and then falls in the tailwater. The falling portion of the lower nappe was considered more significant until it was realized that the dominant low pressures on the corresponding standard overflow occur in the *rising nappe* portion.

a)

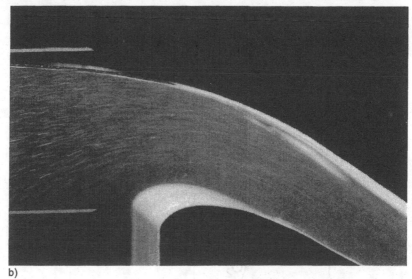

b)

Figure 2.15
Photographs of
(a) sharp-crested
weir flow,
(b) standard-
crested overflow at
design flow

For overflow depths larger than 100 mm, say, where so-called scale effects are absent, the weir flow and the overflow over a standard crest are fully equivalent. For a weir crest located by an amount Δz over the sharp crest, equal discharge passes for an equal upstream level. The shape of the crest is important regarding the *bottom pressure distribution*. Slight modifications have a significant effect on the bottom pressure, while the discharge characteristics remain practically the same. The geometry of the lower nappe cannot simply be expressed analytically. The best known approximation is due to the US Corps of Engineers (USCE, 1970). They proposed a three arc profile for the upstream quadrant and a power function for the downstream quadrant, with the crest as origin of the Cartesian coordinate system $(x; z)$. The crest shape is plotted in Figure 2.16.

The significant scaling length for the standard overflow structure is the so-called *design head* H_D (subscript D for normal design). All other lengths may be non-dimensionalized with H_D, such as the radii of the upstream crest profile $R_1/H_D = 0.50, R_2/H_D = 0.20$ and $R_3/H_D = 0.04$. The origins of curvature O_1, O_2 and O_3, as well as the transition points P_1, P_2, and P_3 for the *upstream quadrant* are detailed in Table 2.1.

Table.2.1 Coordinates of origins of curvature O_i, and transition points P_i for standard crest shape (USCE, 1970).

point	O_1	O_2	O_3	P_1	P_2	P_3
x/H_D	0.000	−0.105	−0.242	−0.175	−0.276	−0.2818
z/H_D	0.500	0.219	0.136	0.032	0.115	0.1360

The *downstream quadrant* crest shape was originally proposed by Craeger (for a later version see Craeger, et al., 1945) as

$$z/H_D = 0.50(x/H_D)^{1.85}, \quad \text{for } x > 0. \tag{2.1}$$

This shape is used up to the so-called tangency-point with a transition to the straight-crested spillway.

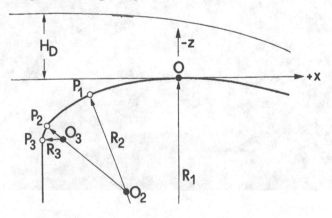

Figure 2.16
USCE crest shape for vertical upstream abutment, and zero velocity of approach

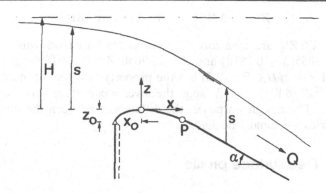

Figure 2.17
Continuous crest
profile with transition
to straight spillway
channel, $P =$
tangency point

The disadvantage of the USCE crest shape is the abrupt change of curvature at locations P_1 to P_3, and at the origin. Such a crest geometry cannot be used for computational approaches due to the curvature discontinuities. An alternative approach with a smooth curvature was provided by Hager (1987). Based on the results of various observations, the crest shape proposed was

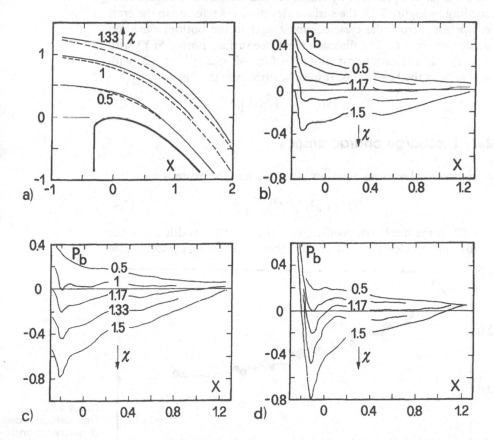

Figure 2.18 Standard overflow structure (a) free surface profile (–) plane flow, (- - -) between piers; bottom pressure distribution for (b) plane flow, (c) axial between piers, (d) along piers (USCE, 1970)

$$Z^* = -X^*\ln X^*, \quad \text{for } X^* > -0.2818 \tag{2.2}$$

where $(X^*; Z^*)$ are transformed coordinates based on Table 2.1 as $X^* = 1.3055(X + 0.2818)$ and $Z^* = 2.7050(Z + 0.1360)$, with $X = x/H_D$ and $Z = z/H_D$. Eq. (2.2) has the property that the second derivative is $d^2 Z^*/dX^{*2} = -1/X^*$, e.g. the inverse curvature varies linearly with X^*. For design purposes, the differences between the two crest geometries are usually negligible.

2.2.3 Free surface profile

The free surface over a standard overflow structure is important in relation to freeboard design and for gated flow. Figure 2.18(a) refers to the USCE curves both for 2D-flow and for the axial profiles between two piers. Along the crest piers, the free surface is lower than for plane flow due to the transverse acceleration.

A *generalised approach* for the plane flow over the standard-shaped overflow crest was provided by Hager (1991a) as shown in Figure 2.19. According to Figure 2.17, the surface elevation s is referred to the crest level upstream from the crest origin O, and to the bottom elevation downstream from O. The dimensionless free surface profile $S(X)$ with $S = s/H_D$ decreases almost linearly for $-2 < X/\chi^{1.1} < +2$ with $\chi = H/H_D$ as the head ratio. It may be approximated as

$$S = 0.75[\chi^{1.1} - (1/6)X]. \tag{2.3}$$

2.2.4 Discharge characteristics

The discharge Q over an overflow structure may be expressed as

$$Q = C_d b (2gH^3)^{1/2} \tag{2.4}$$

where C_d is the discharge coefficient, b the overflow width and g the gravitational acceleration. As all the other parameters, C_d varies also

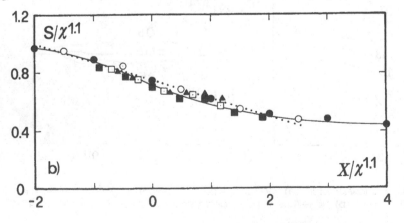

Figure 2.19
Free surface profile, experiments and (–) average data curve, (···) Eq.(2.3)

with the relative head $\chi = H/H_D$ only. The experimental data plotted in Figure 2.20(b) may be approximated independently of the spillway angle α up to $\chi = 3$ as (Figure 2.17)

$$C_d = \frac{2}{3\sqrt{3}}\left[1 + \frac{4\chi}{9 + 5\chi}\right]. \tag{2.5}$$

For $\chi \to 0$, the overflow is shallow and almost hydrostatic pressure occurs. Then, the overflow depth is equal to the critical depth, and the discharge coefficient is $C_d = 2/(3\sqrt{3}) = 0.385$. For design flow, $\chi = 1$ as discussed below, and $C_d = 0.495$. The discharge coefficient may thus be increased by 30%. For $\chi > 2$, the increase is only moderate, such as $C_d(2) = 0.55$.

2.2.5 Bottom pressure characteristics

The bottom pressure distribution $p_b(x)$ is important because it yields:

- an index for the potential danger of cavitation damage, and

- the location where piers can end without inducing separation of flow.

Figure 2.18(b) to (d) provide bottom pressure head data $p_b/(\rho g)$ nondimensionalized by the design head H_D as $P_b = p_b/(\rho g H_D)$, as a function of location $X = x/H_D$ for various values of χ. It is noted that the *minimum* (subscript m) pressure p_m occurs on the upstream quadrant throughout, and that bottom pressures are positive for $\chi = H/H_D \le 1$. The most severe pressure minima occur along the piers due to significant streamline curvature effects.

A generalized analysis of data was conducted by Hager (1991a) by accounting for all published observations relative to *plane* overflow. The data are plotted as $\bar{P}_m = p_m/(\rho g H)$ over the relative head χ on the overflow structure in Figure 2.20(a) and they may be approximated as

$$\bar{P}_m = (1 - \chi). \tag{2.6}$$

Accordingly, the minimum bottom pressure is positive compared to the atmospheric pressure when $\chi < 1$. Also, the minimum pressure head $p_m/(\rho g)$ is proportional to the effective head H, and $1 - \chi$. The location of minimum pressure is $X_m = -0.15$ for $\chi < 1.5$, and $X_m = -0.27$ for $\chi > 1.5$, i.e. just at the transition of the crest to the vertical abutment.

The crest (subscript c) bottom pressure index $\bar{P}_c = p_c/(\rho g H)$ as a function of χ is seen to be significantly above the minimum bottom pressure (Figure 2.21(a). Approximately, the observations yield $\bar{P}_c = (2/3)\bar{P}_m$. Figure 2.21(b) refers to the location of zero bottom pressure, i.e. where atmospheric pressure occurs. The data vary

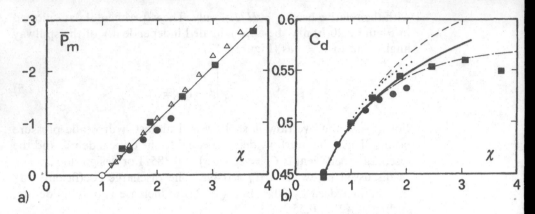

Figure 2.20 (a) Minimum bottom pressure index $\bar{P}_m = p_b/(\rho g H)$ and (b) discharge coefficient C_d as functions of relative head $\chi = H/H_D$

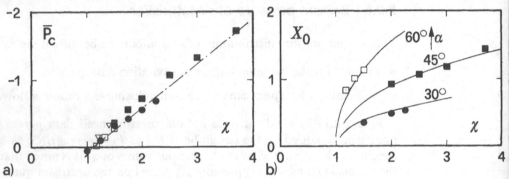

Figure 2.21 (a) Crest pressure $\bar{P}_c(\chi)$, (b) location of atmospheric bottom pressure $X_0(\chi)$

not only with the relative head χ but also with the tailwater chute angle α. The relative position $X_0 = x_0/H_D$ may be given as (Hager, 1991a)

$$X_0 = 0.9\tan\alpha(\chi - 1)^{0.43}. \tag{2.7}$$

2.2.6 Velocity distribution

Velocity distributions were recorded by Hager (1991a, b) for spillway angles $\alpha = 30°$ and $45°$. Figure 2.22 shows normalized plots for the relative velocity $\mu = V/(2gH_D)^{1/2}$ as functions of X and Z for various relative heads χ. Away from the crest, the velocity distribution is almost uniform and the pressure distribution is thus almost hydro-static. However, in the crest domain, there is a significant increase of velocity in the streamwise and depthwise directions. This must be attributed to the free surface gradient and the streamline curvature, respectively. Due to the generalized plots, one may read off velocities at any location $(X; Z)$ from Figure 2.22.

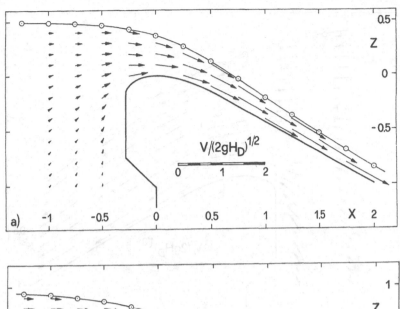

Figure 2.22
Velocity
distribution
$V/(2gH_D)^{1/2}$ as
functions of $(X;Z)$
for $\chi = $ (a) 0.5, (b)
1, (c) 1.5, and (d) 2

Figure 2.23 refers to typical plane flow over a standard overflow structure. The flow is seen to be absolutely smooth, and small air bubbles contained in the approach flow reveal the streamlines. The design head H_D of the structure tested was $H_D = 0.20\,\mathrm{m}$ and is indicated with white lines. The tailwater bottom angle was $\alpha = 30°$.

The *crest velocity distribution* $V_c(z)$ is of particular relevance and Figure 2.24(a) shows the distributions normalized by H_D as $\mu_c = V_c/(2gH_D)^{1/2}$. The increase of crest velocity with both depth z and χ is obvious, except for the thin bottom boundary layer. An even simpler plot yields a normalization by the effective head H, instead of

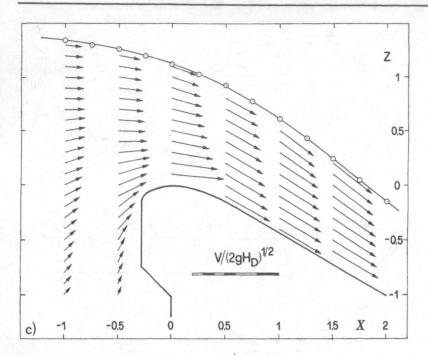

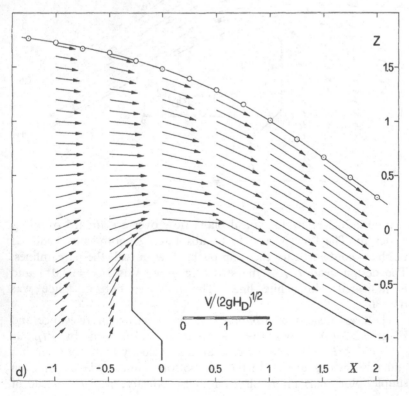

Figure 2.22
(Continued)

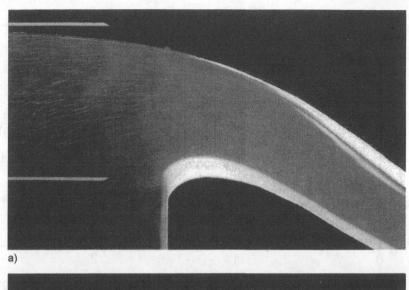

a)

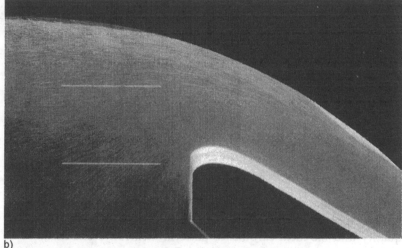

b)

Figure 2.23
Flow pattern at
standard overflow
structure for $\chi =$
(a) 1, (b) 2

the design head, H_D. Computations reveal that the *free vortex*
model approximates the crest section favorably, i.e. that the product of tangential velocity times the radius of curvature
remains constant. The resulting velocity distribution thus is (Hager,
1991b)

$$\frac{V_c}{(2gH)^{1/2}} = \frac{1}{2}\left[\frac{r+0.75}{r+z/H}\right] \tag{2.8}$$

where $r = d^2Z/dX^2 = 0.584$ is the dimensionless crest curvature from
Eq.(2.2). A complex plane flow problem is thus governed by a fundamental physical law.

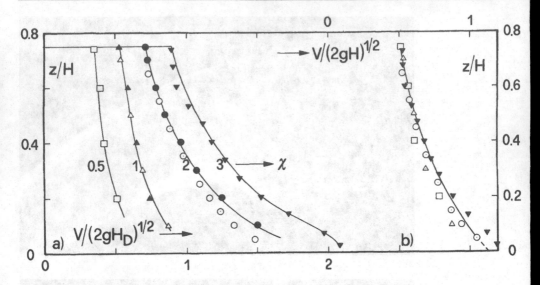

Figure 2.24 Crest velocity distribution (a) $V_o/(2gH_D)^{1/2}$ and (b) $V_o/(2gH)^{1/2}$ as functions of relative elevation over the crest z/H for various χ

2.2.7 Cavitation design

Standard overflows with $\chi < 1$ are referred to as underdesigned, while an *overdesign* involves $\chi > 1$, and thus subatmospheric bottom pressures. Initially, overdesign of dam overflows was associated with advantages in capacity. However, as follows from Figure 2.20(b) the increase of C_d for $\chi > 1$ is relatively small, but the decrease of minimum pressure p_m from Figure 2.20(a) is significant. Overdesigning does thus add to the cavitation potential. *Incipient cavitation* is known to be a statistical process depending greatly on the water quality and the local turbulence pattern. Generally, one assumes an incipient pressure head $p_{vi}/(\rho g) = -7.6$ m as compared to -10 m roughly for static water (Abecasis, 1970). The limit (subscript L) head H_L for incipient cavitation to occur thus is from Eq.(2.6)

$$H_L = [\gamma(1 - \chi)]^{-1}[p_{vi}/(\rho g)]. \tag{2.9}$$

Because the incipient cavitation head is a length, H_L is also expressed in [m]. Figure 2.25 compares Eq.(2.9) with data from Abecasis and indicates general agreement. The constant γ was introduced to account for additional effects, such as the variability of p_{vi} with χ.

2.2.8 Overflow piers

Piers on overflow structures are provided:

Figure 2.25
Limit head H_L(m) as
a function of
relative head
$\chi = H/H_D$ on plane
overflow structure
(–) Eq.(2.9), ($\cdots$)
domain of
Abecasis, ©
domain of definite
cavitation

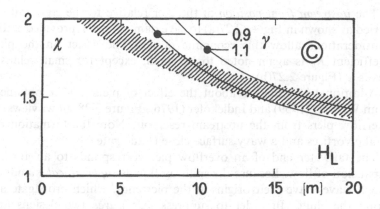

- to improve the approach conditions,
- to mount overflow gates,
- to divide the spillway in various chute portions, and
- to aerate the chute flow at the pier ends.

The *front pier shape* was studied by the USCE, and two typical designs are shown in Figure 2.26(a). Other front shapes including an almost rectangular pier nose with rounded edges, and a triangular nose were also proposed, but the designs shown are standard.

A pier modifies the overflow from plane to spatial. In terms of discharge, this effect is accounted for by the *effective width* b_e instead of the geometrical width b as

$$b_e = b - 2K_pH \tag{2.10}$$

where K_p is a pier coefficient, and H the head on the overflow structure. The parameter K_p decreases according to Figure 2.26(b) slightly from 0.05 for $\chi = 0.2$ to $K_p = -0.05$ for $\chi = 1.5$, but may be assumed almost equal to $K_p = 0$ for preliminary design purposes. Accordingly, the recommended piers do not significantly perturb the plane flow pattern.

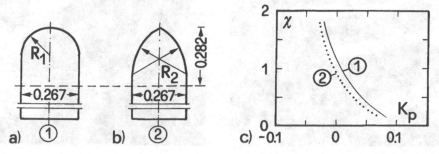

Figure 2.26 Typical pier front shapes (a) circular, (b) circular-arced, with numbers to be multiplied with design head H_D, (c) K_p values according to USCE (1970)

The *upstream front position* of the pier relative to the crest can be varied as shown in Figure 2.27(a) to save dam material, provided static considerations allow this economic design. The effect on the pier coefficient K_p is again noted to be small, except for small relative heads χ (Figure 2.27(b)).

Additional information about the effect of piers can be obtained from Webster (1959) and Indlekofer (1976). Figure 2.28 shows views of overflow piers from the upstream reservoir. Note the formation of intake vortices and a wavy surface close to the gates.

The tailwater end of an overflow pier corresponds to an abrupt expansion of flow. Because the spillway flow is *supercritical*, standing shock waves have their origins at the pier ends, which propagate all along the chute. In order to suppress *pier waves* two designs are available:

- either sharpening the pier end both in width and height, or

- continue with the pier as a dividing wall along the chute.

Both designs are not ideal, because even a slim pier end perturbs the flow and dividing walls may be costly especially for long spillways.

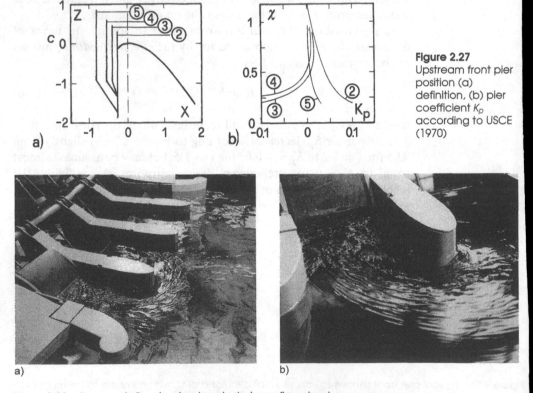

a)   b)

Figure 2.27 Upstream front pier position (a) definition, (b) pier coefficient K_p according to USCE (1970)

a)   b)

Figure 2.28 Approach flow to piers located at overflow structures

Reinauer and Hager (1994) have proposed an alternative design sketched in Figure 2.29. It involves the so-called *pier extension* (Figure 2.29(c)) corresponding to a one-sided extension wall of an abruptly terminating pier. The shock wave is reduced by the interference principle. The location of the pier extension x_E in the horizontal chute was obtained as

$$X_E = \frac{x_E}{b_p \mathbf{F}_o} = 0.41(2h_o/b_p)^{2/3} \tag{2.11}$$

where b_p is the pier width, $\mathbf{F}_o = V_o/(gh_o)^{1/2}$ the approach Froude number with V_o as approach velocity at the pier end, and h_o as corresponding flow depth.

For spillway flow without pier extension, a pier wave 1 in the pier axis, and a reflection wave 2 from the adjacent pier, or by the presence of a chute wall occur. Both waves depend significantly on the ratio $(h_o/b_p)^{1/2}$, as found by Reinauer and Hager (1994). The pier extension may be considered a simple means to save costs for high spillway walls, and to inhibit uncontrolled aeration at the pier ends.

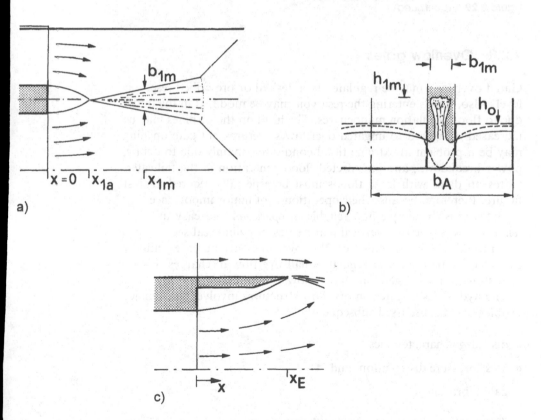

Figure 2.29 Pier wave (a) plan view, (b) section for untreated flow, (c) effect of pier extension, (d) photograph

d)

Figure 2.29   *(Continued)*

2.2.9   Overflow gates

Gated overflows may be regulated to a desired or prescribed reservoir level. Also, floods entering the reservoir may be modified in the outlet due to flood regulation manoeuvres. The head on the turbines may be increased compared to ungated overflow structures but gate opening may be a problem in extreme flood conditions, mainly due to debris, power breakdown, and unexpected flood generation in the tailwater. Gates on dams with large floods must be especially secured against failure, therefore, because their operation is of major importance.

Gates must be simple and reliable in *operation* and easy in *maintenance*. Usually, either vertical plane gates or cylindrical sector gates are provided on the crest of the overflow structure, eventually extended by flap gates to regulate small overflow discharges. Figure 2.30 shows a typical gate on an overflow structure.

The hydraulics of gates on overflow structures involves three major problems to be discussed subsequently:

- discharge characteristics,

- crest pressure distribution, and

- gate vibration.

The *vertical plane gate* located at the crest section as shown in Figure 2.31(a) was analysed by Hager and Bremen (1988). The discharge Q

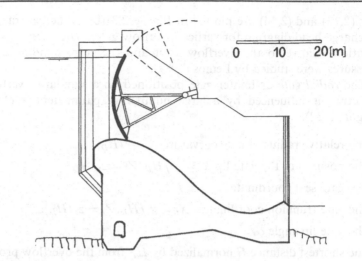

Figure 2.30
Sector gate on an
overflow structure
with a standard
crest

depends on the head H_o on the gate, the corresponding head H without
the gate, the gate opening a, and the crest design head H_D. The design
discharge Q_D is defined as

$$Q_D = C_{dD}b(2gH_D^3)^{1/2} \tag{2.12}$$

where $C_{dD} = C_d(\chi = 1) = 0.495$ from Eq.(2.5). The discharge Q_g
under a gated overflow was shown to be

$$Q_g/Q_D = [\chi_o^{3/2} - (\chi_o - A)^{3/2}][(1/6) + A]^{1/9} \tag{2.13}$$

where $\chi_o = H_o/H_D$ is the relative head of gated flow and $A = a/H_D$ is
the relative gate opening. Eq.(2.13) holds for $0 < A < 2$ and $\chi_o >
(4/3)A$ because the gate is not submerged for $\chi_o < (4/3)A$. Then, the
overflow is ungated, i.e. $\chi_o \rightarrow \chi$, and $Q_g \rightarrow Q$, for which Eq.(2.13)
simplifies to

$$Q/Q_D = \frac{2}{3\sqrt{3}} \frac{\chi^{3/2}}{C_{dD}} \left[1 + \frac{4\chi}{9 + 5\chi}\right]. \tag{2.14}$$

Figure 2.31
Vertical gate at
crest of overflow
structure
(a) definition,
(b) discharge-head
equation (–)
Eq.(2.13), (···)
Eq.(2.14) for
ungated flow

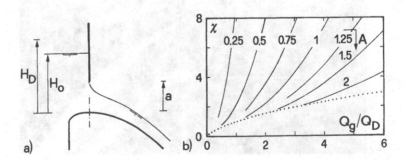

Eqs.(2.13) and (2.14) are plotted in Figure 2.31(b) as the generalized
discharge head diagram for vertical overflow gates. The effect of gate
location relative to the overflow crest and the corresponding crest
pressures were studied by Lemos (1981).

The *radial gate* or tainter gate positioned on a standard overflow
structure is influenced by a large number of parameters, such as
(Figure 2.32):

- the relative radius of gate curvature $R = r/H_D$,
- the position of the gate lip $X_l = x_l/H_D; Z_l = z_l/H_D)$,
- the gate seat coordinate $X_s = x_s/H_D$,
- the gate trunnion coordinates $(X_t = x_t/H_D; Z_t = z_t/H_D)$,
- the gate lip angle α,
- the shortest distance G normalized by H_D from the overflow profile
  to the gate lip,
- the corresponding horizontal coordinate $X_w = x_w/H_D$, and
- the profile angle γ at this point.

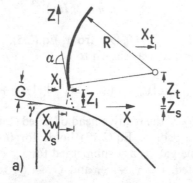

a)

b)

Figure 2.32
Radial gate on
standard overflow
structure
(a) definition of
parameters,
(b) overflow photo

The independent parameters are the design head H_D of the overflow crest, the operational head H, and the effective head H_e corresponds to the approach head over the crest for $x_l < 0$, and to the approach head plus the height of the standard profile for $x_l > 0$. The discharge may be computed from

$$Q = C_{dg}b(GH_D)(2gH_e)^{1/2} \tag{2.15}$$

where C_{dg} is the discharge coefficient for gated flow. Sinniger and Hager (1989) have presented a complete analysis with which the discharge characteristics may be obtained.

According to Rhone (1959) the gate seat coordinate should be positive to inhibit significant underpressure, but x_s should be confined to small values to limit the height of gate. Typically, a value between $0 \le X_s \le +0.2$ should be chosen, for which Z_s is smaller than 0.025.

The *bottom pressure distribution* over a gated standard overflow structure was also observed by Lemos (1981), among others. The location of minimum pressure was always about $\Delta X = +0.20$ downstream from the gate seat coordinate. These subpressures are determined for both cavitation and separation from the overflow profile.

The *minimum bottom pressures* $P_m = p_m/(\rho g H_D)$ are practically uninfluenced by the relative gate radius R, whereas the effect of χ_o is significant. The curves $P_m(Z_l)$ increase typically to a maximum value at $Z_l = 0.4$ which is equal to roughly $P_m = -0.2$ for $\chi = 1$, and $P_m = -0.4$ for $\chi = 1.25$, depending on the trunnion elevation. For a detailed analysis, refer to Lemos (1981) or Sinniger and Hager (1989).

2.3   SIDE CHANNEL

2.3.1   Typology

The side channel or the side channel spillway is a common structure used for overflows. The axis of the side channel is parallel to the overflow crest, whereas the axis is perpendicular to the crest in frontal overflow structures, as discussed in 2.2. Whereas a frontal overflow is located at the dam structure, the side channel is separated from the dam, and the discharge is conveyed along the valley down to the tailwater. Figure 2.33 shows a typical configuration involving the side channel, the spillway and the stilling basin as the dissipator.

Side channels are often considered at sites where:

- a narrow gorge does not allow sufficient width for the frontal overflow,
- impact forces and scour are a problem in case of arch dams,
- a dam spillway is not feasible, such as in the case of an earth dam, or
- topographic conditions are favourable for a side channel.

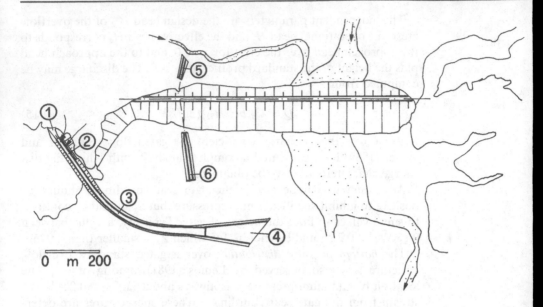

Figure 2.33 General arrangement of side channel. ① Frontal gated inlet, ② side channel, ③ spillway, ④ stilling basin, ⑤ intake structure, ⑥ bottom outlet

Side channels are definitely not considered for dams with a large design flood due to the limited overflow head of say 3 m. Also, side channels are usually not equipped with gates, and *drum gates* are often the best choice for cases in which the level regulation is important. The length of the overflow structure can be increased by discharging from both sides into the side channel. Figure 2.34 refers to a design with multiple inlets.

The *cross-section* of a side channel is either rectangular or, for large overflow structures, trapezoidal to save excavation cost. Side channels

Figure 2.34
Side channel with a multiple discharge inlet (Bretschneider, H. *Wasserwirtschaft* **61**(5): 143)

Figure 2.35 Typical views at side channel (a) upstream and (b) downstream portions, (c) plan with two-cell vorticity flow

should not submerge the reservoir outflow due to complex interactions between the overflow and the downstream channel. Also, no blocks or other appurtenances should be located along the invert of the side channel to still the lateral flow or to direct the overflow in the axis of the side channel (USBR, 1938) because of the axial flow disturbance. The bottom slope of the side channel is usually constant, with a possible exception at the dead end where a transition from the overflow to the side channel invert may be used. The roughness of the side channel is comparable with that of a spillway, i.e. smooth concrete is provided. The width of a side channel is often constant, or slightly

diverging with increasing length but there are no hydraulic advantages
from the latter design regarding the headloss.

2.3.2 Hydraulic design

The one-dimensional equation for the free surface profile can be
derived from momentum considerations (Chow, 1959). If h is the
local flow depth varying with the longitudinal coordinate x, S_o the
bottom slope, S_f the friction slope, $U\cos\phi$ the lateral inflow component in streamwise direction, V the average side channel velocity, Q the
local discharge, g the gravitational acceleration, A the cross-sectional
area with $\partial A/\partial x$ as surface width change, and $\mathbf{F}^2 = Q^2(\partial A/\partial h)/(gA^3)$
the square of the local Froude number, then

$$\frac{dh}{dx} = \frac{S_o - S_f - \left[2 - \frac{U\cos\phi}{V}\right]\frac{Q(dQ/dx)}{gA^2} + \frac{Q^2(\partial A/\partial x)}{gA^3}}{1 - \mathbf{F}^2}. \qquad (2.16)$$

For $dQ/dx = 0$, i.e. no lateral inflow, and $\partial A/\partial x = 0$, i.e. a prismatic
channel, this relation reduces to the backwater equation. Eq.(2.16) can
also be derived from energy considerations to yield (Hager, 1994a)

$$H = h + z + \frac{V^2}{2g}, \quad \frac{dH}{dx} = -\left[S_f + \left(1 - \frac{U\cos\phi}{V}\right)\frac{Q(dQ/dx)}{gA^2}\right] \qquad (2.17)$$

and indicates that the change of total head H with x (where z is the
elevation above a reference datum) is equal to the friction slope S_f plus
a term proportional to the local velocity $V = Q/A$ divided by a velocity $V_d = gA/(dQ/dx)$, times a reduction factor depending on the
lateral inflow characteristics. For an inflow direction of 90°, the latter
term is equal to unity, and the term $V/V_d \geq 0$ must be added to the
friction slope. This term is large for either large side channel velocity or
large lateral inflow.

  If all channel geometry, lateral flow, roughness and a boundary
condition are specified, one may solve Eq.(2.16) numerically. However, this may be simplified for practical purposes as follows. Under
the assumptions of:

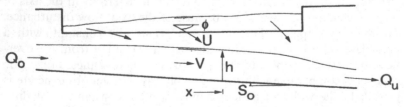

Figure 2.36
Definition of flow in
type side channel

- a prismatic side channel $\partial A/\partial x = 0$,

- an average value $S_{fa} = (S_{fu} + S_{fd})/2$ for the friction slope with subscript u and d for the up- and downstream ends, instead of a spatially variable friction slope $S_f(x)$, and

- a constant lateral inflow intensity $p_s = dQ/dx$

the difference $J = S_o - S_{fa}$ is independent of x and corresponds to a *substitute slope* larger than zero. Using the coordinates

$$x_s = \frac{8p_s^2}{gb^2J^3}, \quad h_s = \frac{4p_s^2}{gb^2J^2} = Jx_s/2 \qquad (2.18)$$

for a typical location x_s, and a typical flow depth h_s, and nondimensionalising as $X = x/x_s$, $y = h/h_s$ yields the governing *free surface equation* instead of Eq.(2.16)

$$\frac{dy}{dx} = 2\frac{y^3 - Xy}{y^3 - X^2}. \qquad (2.19)$$

The scalings $(x_s; h_s)$ have been chosen such that Eq.(2.19) becomes *singular* for $(X; y) = (1; 1)$, i.e. both numerator and denominator tend to zero. In other words, this singularity (subscript s) occurs for a flow which is simultaneously uniform because $dy/dX = 0$, and critical because $dX/dy = 0$. Physically such flow occurs for side channels in which the tailwater flow is supercritical and a *transition* from subcritical upstream (typically at a dead-end) to supercritical downstream flow develops.

For all side channels of length L_s with a supercritical tailwater flow, i.e. where the bottom slope of the downstream channel is larger than the critical slope, one may distinguish between two cases (Figure 2.37):

- if the length L_s is larger than the location of the singular point x_s, then a transition from sub- to supercritical flow occurs at the *singular point* $x = x_s$; and

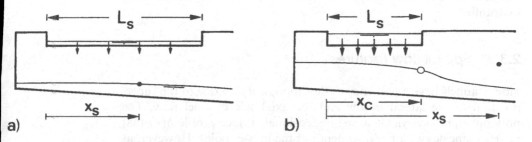

a)   b)

Figure 2.37 Supercritical flow in side channel (a) singular point, (b) critical point

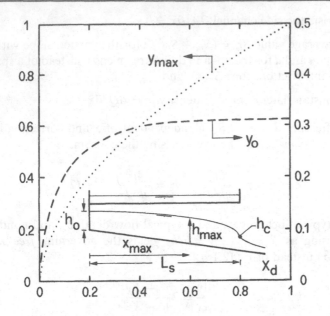

Figure 2.38
Side channel with
critical flow at
downstream end,
upstream flow
depth y_o and
maximum flow
depth y_{max} as a
function of
$X_d = L_s/x_s$.

- if the length L_s is smaller than x_s, the latter has no physical relevance and critical flow is forced at the end of the side channel, at the *critical point $x = L_s$*.

Eq.(2.19) can be solved numerically (Hager, 1994a) but there are two types of information of design relevance in connection with dam side channels where the tailwater is *supercritical* throughout:

- what is the maximum flow depth h_{max}, and where is its location x_{max} along the side channel?, and

- what is the dead-end flow depth h_o?

Figure 2.38 shows both $y_{max} = h_{max}/h_c$, and $y_o = h_o/h_c$ as a function of $X_d = L_s/x_s \le 1$ based on detailed numerical computations. With the critical flow depth $h_c = [Q_d^2/(gb^2)]^{1/3}$, the upstream depth h_o, and the maximum depth h_{max}, an approximate free surface profile may be drawn. The location of maximum depth is $X_{max} = y_{max}^2$ from Eq.(2.19). Flows with subcritical tailwater are not typical in dam hydraulics.

2.3.3 Spatial flow features

Side channels have a tendency to develop *spiral flow* due to the super-position of the lateral inflow and the axial side channel flow. The previous approach yields a one-dimensional surface profile $h(x)$ that nearly coincides with the flow depth at the impact point. However, as shown in Figure 2.39 for various side channel shapes, the wall flow

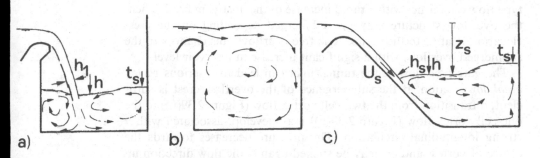

Figure 2.39 Lateral flow configuration in side channel (a) free and (b) submerged flow in rectangular section, (c) free flow in trapezoidal section

depth t_s, or the corresponding axial depth for twin inlets may be considerably larger than $h(x)$ due to spiral currents. According to limited observations, and with $P_h = b + 2h$ as wetted parameter, z_s as fall depth and for a trapezoidal side channel of transverse slope 1 (horizontal) : 0.6 (vertical) one has

$$\frac{t_s}{h} = 1 + 5.5\left[\frac{p_s^2}{ghP_h}(z_s/h)^{\frac{1}{2}}\right]^{1/2}. \tag{2.20}$$

For rectangular side channels with $U_s = p_s/h_s$ as lateral inflow velocity the result is

$$\frac{t_s}{h} = 1 + \gamma_s \frac{p_s U_s}{gh^2} \tag{2.21}$$

where the average proportionality factor was $\gamma_s = 1$, and the maximum could rise to $\gamma_s = 1.5$. The local flow depth $h(x)$ is computed from a 1-D approach, such as previously outlined.

Side channels often discharge into tunnel spillways. In order to satisfy the basic requirements for dam overflow structures, such tunnels must always have a *free surface flow*, because submergence can easily lead to dam overtopping. Of particular concern is the tunnel inlet as shown in Figure 2.40. If the flow touches the inlet vertex a gate

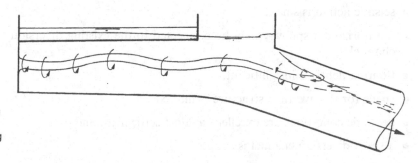

Figure 2.40
Side channel with a tailwater spillway tunnel under surging flow

type flow establishes with a rapid increase of the upstream head. Then, the overflow structure may get submerged depending on the crest elevation relative to the vertex. The free overflow thus changes to the submerged overflow, with a significant increase in reservoir level.

The gate type flow at the tunnel inlet can lead to a serious *vortex breakdown* action. If the submergence of the overflow crest is sufficient, a transition from the two-cell vortex flow (Figure 2.39(a)) to the one-cell vortex flow (Figure 2.39(b)) may develop, associated with a strong longitudinal circulation. The pressure decreases towards the centre of vortex, and air may be sucked against the flow direction up to the dead-end, where the line vortex attaches to the wall. Such a vortex breakdown may seriously decrease the discharge capacity and lead to an additional reservoir level rise. It is thus important to verify free flow not only for the side channel, but also for a tailwater spillway surface tunnel.

2.4 MORNING GLORY OVERFALL

2.4.1 Concept

A shaft type spillway or Morning Glory overflow structure is composed of three components, namely the cup-shaped overflow, the vertical shaft, and the nearly horizontal diversion spillway. The design discharge has to produce a *free surface flow*, i.e. both the shaft and the diversion spillway must have a larger capacity than the overflow structure. In order to promote atmospheric pressure all along the spillway up to the dissipator, an aeration conduit is normally inserted in the shaft. A typical section of the structure is shown in Figure 2.4(c).

Morning Glory spillways are typically used for dams with small to medium design discharges, with a maximum of approximately $1000 \, \text{m}^3 \, \text{s}^{-1}$. The overfall height of the structure may be almost 100 m although 50 m is more relevant. The structure has a circular standard-crested overfall, a vertical shaft, a bottom bend including the aeration device and a diversion tunnel discharging into the energy dissipator. This structure is advantageous when:

- seismic action is small,
- the horizontal spillway may be connected to the existing diversion channel,
- floating debris is insignificant,
- space for the overflow structure is limited,
- geologic conditions are excellent against settlement, and
- a short diversion channel is sought.

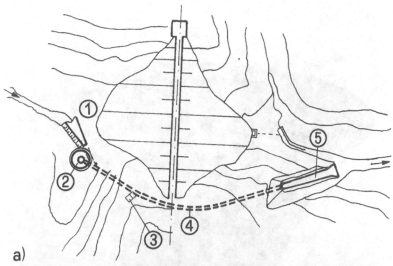

a)

Figure 2.41
(a) Morning glory
type structure with
① intake for river
diversion during
dam construction,
② overflow structure,
③ inspection entry,
④ tunnel spillway,
⑤ stilling basin; (b) El
Makhazine dam on
Oued Loukkos
(Morocco)
including overflow,
intake and bottom
outlet structures. The
structure is 50 m high
and designed for
1450 m³ s⁻¹

b)

Debris may not be a concern if the crest and shaft radius are sufficiently large (see below). Figure 2.41 shows a typical arrangement of
the type structure in a project involving an earth dam.

The design of the Morning Glory spillway involves the overflow
structure, the shaft structure, the tunnel spillway and the aeration
conduit. The latter item was poorly studied up till now, and a hydraulic
model study is recommended for this type of spillway structure. Also,
the approach currents to the structure should be studied using a
general model to account for the spatial flow features. The structure
is prone to rotational approach flow, which should be inhibited with a
selected location of the shaft relative to the reservoir topography and
the dam axis. The *radial flow* may be improved with piers positioned

Figure 2.42
Morning glory
spillway of Genzano
dam (Italy), view on
single intake (Salini
Costruttori S.p.A.
Italy)

on the overfall crest. Figure 2.42 relates to a typical Morning glory
overflow.

2.4.2 Crest shape

The shape of the Morning Glory overfall is a logical extension of the
standard overfall crest. In order to control the crest pressures, sharp-
crested circular weirs were tested and both the discharge characteristics
and the lower crest geometry were experimentally determined. A basic
study was conducted by Wagner (1956) as an extension of the 1948

USBR project. The experimental arrangement is shown in Figure 2.43(a). All quantities referring to the circular sharp-crested weir are overbarred to indicate the difference with the standard overflow structure. The pipe radius is $\bar{R}$, the overflow head relative to the sharp crest is $\bar{H}$ and the coordinate system $(\bar{x}; \bar{z})$ is located at the weir crest.

As for the overflow with a straight crest in plan view, the circular weir overflow has a lower nappe increasing to a *maximum* (subscript *m*) point P with coordinates $(x_m; z_m)$. The coordinate system of the standard Morning Glory overflow is located at P, the corresponding crest radius is $R = \bar{R} - x_m$, and the head is $H = \bar{H} - z_m$.

If $\bar{X} = \bar{x}/\bar{H}$ and $\bar{Z} = \bar{z}/\bar{H}$ are the dimensionless nappe coordinates the location of point P is for $0.1 < \bar{H}/\bar{R} < 0.5$ (Hager, 1990)

$$\alpha = \bar{X}_m/\bar{X}_{mo} = [1.04 - 1.055(\bar{H}/\bar{R})], \tag{2.22}$$

$$\beta = \bar{Z}_m/\bar{Z}_{mo} = [1.04 - 1.020(\bar{H}/\bar{R})]. \tag{2.23}$$

Note the differences of 4% for $\bar{H}/\bar{R} \to 0$ from the observations $(\bar{X}_{mo}; \bar{Z}_{mo}) = (0.250; 0.112)$.

The *nappe profiles* are shown in Figure 2.44(a) for various relative heads $\bar{H}/\bar{R}$. The overflow is limited to $\bar{H}/\bar{R} < 0.40$. For larger relative heads, the pipe is submerged and the overflow changes to orifice flow. According to Wagner (1956), the effect of approach velocity is negligible for $\bar{w} > 2\bar{R}$ where $\bar{w}$ is the weir height. Scale effects are insignificant for heads $\bar{H}$ larger than 40 mm.

According to Figure 2.44(a) a distinct nappe profile may be attributed to each value of $\bar{H}/\bar{R}$. However, when using the transformations $\bar{X}^* = \bar{X}/\alpha$ and $\bar{Z}^* = \bar{Z}/\beta$ where α and β are defined in Eqs.(2.22) and (2.23), a single generalized nappe profile for all relevant relative radii may be obtained (Figure 2.44(b)). The equation of the lower nappe thus reads for $X^* < 1.6$

$$Z^* = -X^* \ln X^* \tag{2.24}$$

in analogy to Eq.(2.2) and the *upper nappe* may be approximated as

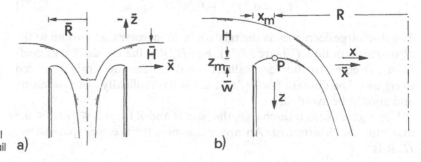

Figure 2.43
Crest geometry of sharp-crested circular pipe overflow (a) overall view, (b) crest detail a)   b)

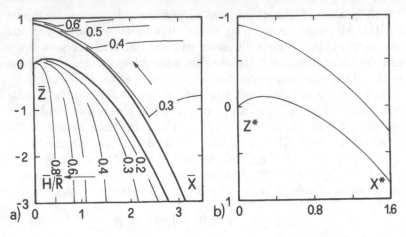

Figure 2.44
Nappe profiles for various relative pipe radii $\bar{H}/\bar{R}$ (a) $\bar{Z}(\bar{X})$ and (b) generalisation $Z^*(X^*)$ for $0.2 \leq \bar{H}/\bar{R} < 0.6$, $(-)\bar{H}/\bar{R} = 0$ (Hager, 1990)

$$\bar{Z} = 1 - 0.26 \left(\frac{\bar{X}}{1 - 1.2(\bar{H}/\bar{R})^{2.3}} + 0.6 \right)^2 . \qquad (2.25)$$

It would be possible to transform these relations to the (X, Z) coordinate system of the overfall structure, but it is easier to refer to the coordinates $(\bar{X}; \bar{Z})$ of the corresponding sharp-crested weir. Once the basic parameters $\bar{H}$ and $\bar{R}$ are determined, this may easily be accomplished.

2.4.3 Discharge and pressure characteristics

The discharge over a Morning Glory overfall structure is in analogy with the straight-crested overfall

$$Q = C_d 2\pi R (2gH^3)^{1/2} \qquad (2.26)$$

where C_d is the discharge coefficient relative to the parameter R and H (Figure 2.43(b)). According to Indlekofer (1978) and for values $0.2 < H/R < 0.5$ (Figure 2.45(b))

$$C_d = 0.515[1 - 0.20(H/R)]. \qquad (2.27)$$

The discharge decreases as the relative head increases, according to the obstruction of flow (Figure 2.44a). For $H/R = 0.2$ the discharge coefficient is equal to the base value $C_d = 0.495$ for the straight-crested overflow. The domain $0 < H/R < 0.2$ is hydraulically not interesting and should be avoided.

For a given design discharge, the pair H and R for $0.2 \leq H/R \leq 0.5$ may thus be determined. An approximation for the weir parameters $\bar{H}, \bar{R}$ is

$$H/R = 1.06(\bar{H}/\bar{R})^{1.07}. \qquad (2.28)$$

Then, the crest coordinates may be determined from (2.22) and (2.23), and the nappe profiles are obtained from Figure 2.44.

The *crest pressure* on Morning Glory overfalls was determined by Lazzari in 1955 and presented by Sinniger and Hager (1989). For overflow heads below the design head no underpressures were observed in the crest domain. Attention should be paid to the transition from the crest to the shaft profile that the change of curvature remains small because otherwise the nappe may separate from the bottom. According to Bradley (1956) the crest and the transition to the shaft as well as the difficult transitions along the bottom bend should have an extremely smooth *lining* without offsets and fins to counter cavitation damage.

2.4.4 Location of overflow structure

The Morning Glory overflow structure should be located such that the structural height in the reservoir becomes small and the approach flow remains nearly free of circulation. Also the connection to an existing diversion tunnel should involve the least cost. The *asymmetry* in approach flow may be improved by excavation of protruding soil and rock formations and by providing the overflow with *crest piers*. Figure 2.46(a) refers to an improved approach current of the Lower Shing Mun reservoir in Hong Kong (Smith, 1966). According to detailed model observations, the overflow was improved with a relatively modest excavation. Figure 2.46(b) shows the overflow for 425m^3 s^{-1}. The approach flow was improved by the addition of two piers.

The shape of the reservoir banks around the shaft to improve the radial approach flow was also studied by Novak and Cabelka (1981). The height of the protruding shaft portion from the excavation level to the crest should be at least 1.6 times the design head. The plan shape of excavation was also specified.

The effect of *pier shape* and geometry was analysed by Indlekofer (1976). He proposed parallel-sided or conical piers with a circular-shaped front face. The effect of the pier is normally negligible for free overfall and the information presented for straight-crested overflow structures may be used also for the Morning Glory overflow.

2.4.5 Vertical shaft

The Morning Glory overflow structure is connected with the vertical shaft to the diversion tunnel. The shaft should have a *vertical axis*, a constant diameter D_s and should convey the water along the shaft wall with an internal air flow to guarantee *free surface flow*. If the air flow is cut, the air transport is choked and significant vibrations associated

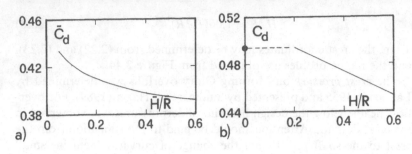

a)

b)

Figure 2.45
Discharge
coefficient (a)
$\bar{C}_d(\bar{H}/\bar{R})$ and (b)
$C_d(H/R)$

with air back-flow may result, and underpressure may develop (Peterka, 1956). To enhance and stabilize the air flow, a separate *aeration conduit* is normally inserted just upstream from the 90° bottom bend. Free surface flow is thus forced also across the bend into the diversion tunnel. Figure 2.47 shows a typical Morning Glory overflow structure with

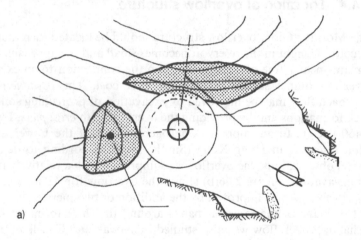

a)

b)

Figure 2.46 Lower Shing Mun reservoir (Hong Kong) (a) Excavation works for improvement of approach flow, (b) overflow patterns (Smith 1966). Submerged flow as shown in Fig 2.46(b) on the right is definitely not recommended

details of the aeration arrangement. Note the offset at the junction between the shaft and the aeration conduit to obtain a stable wall separation. This device is particularly useful for overflows where vibrations are a concern. A less sophisticated aeration uses an aeration conduit not integrated into the overflow structure and discharging sideways into the flow, such as shown in Figure 2.4(c).

For a preliminary design, the shaft radius R_s can be correlated to the crest radius R as

$$R_s[\text{m}] = 1[\text{m}] + 0.1R[\text{m}]. \tag{2.29}$$

A minimum shaft diameter of $R_s = 1.5\,\text{m}$ should be provided, if no surface debris is expected. As surface racks should not be mounted, the shaft has to be sufficiently large to receive logs of wood. If much surface debris must be expected, the Morning Glory overflow is a poor design. The *bend radius* R_b should at least be equal to $6R_s$ for small fall heights up to 20 m, and $10R_s$ for larger fall heights up to 50 m. A number of further structural details are discussed by Bretschneider

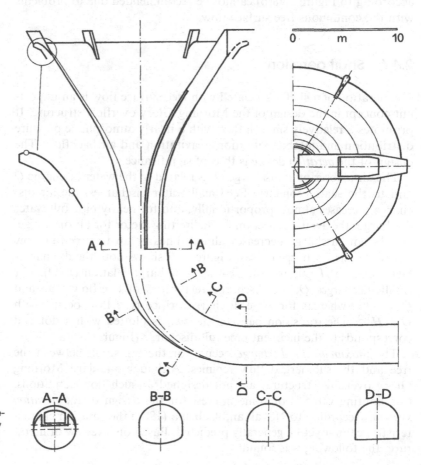

Figure 2.47
Morning Glory
overflow at Innerste-
Talsperre, Germany
(Bretschneider and
Krause, 1965)

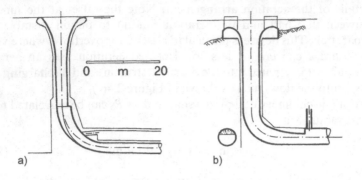

Figure 2.48
Detail of shaft bend,
abrupt expansion at
(a) bend beginning
(Aabach dam,
Germany), (b) bend
end, according to
Gardel (1949), not
recommended

(1980). Currently, no standard shaft design is available, and the shaft bend can be arranged depending on local conditions. It is recommended that the aeration conduit should enter the shaft bend at the *beginning* rather than at its end, as is also shown in Figures 2.48(a) and 2.4(c). This arrangement forces a free surface flow not only in the horizontal diversion tunnel but also along the bend. The design according to Figure 2.48(b) cannot be recommended due to problems with the continuous free surface flow.

2.4.6 Shaft aeration

The aeration of a shaft associated with free surface flow is an important concept in the design of the Morning Glory overflow structure. It promotes a relatively smooth flow with a nearly atmospheric pressure distribution and inhibits vibrations, cavitation and air backflow. The design of the *aeration device* is thus of significance.

The air (subscript a) discharge Q_a is related to the water discharge Q and to the shaft geometry. For small water discharges, the air discharge increases almost proportionally, and for nearly pipe-full water discharge, the available section is the limiting factor for air discharge, i.e. the air discharge decreases almost linearly up to pipe-full flow where the air discharge ceases. Figure 2.49 shows both the discharge-head relation $H(Q)$, and the air-water-discharge relation $Q_a(Q)$. For small discharges Q, the discharge relation is of overflow type and $Q \sim H^{3/2}$, whereas for large discharge orifice type flow occurs with $Q \sim H^{1/2}$. The transition between the two is plotted with a dot and corresponds to the incipient pipe full-discharge (Figure 2.49(a)).

The *maximum air discharge* occurs for the transition between the free and the submerged flow regimes. Although standard Morning Glory overflow structures are not designed to such flow conditions, this limiting value is of some interest for the design of the *aeration system*. According to Novak and Cabelka (1981) the aeration characteristics cannot yet be generally predicted. Based on a review of literature, the following was found:

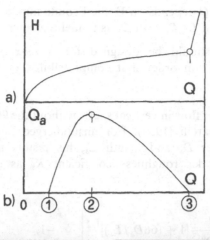

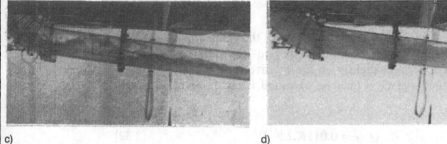

Figure 2.49 Discharge characteristics of shaft spillways (a) head-discharge relation, (b) air-water discharge relation, (c) unsubmerged and (d) submerged shaft flow for final design of Lower Shing Mun reservoir (Hong Kong)

- The aeration of a Morning Glory overflow depends mainly on the shaft geometry and the bend shape,

- The maximum air discharge occurs for supercritical tunnel flow, i.e. without a hydraulic jump downstream from the shaft bend;

- By assuming co-current air-water tunnel flow, and with A_t as tunnel section and A_w as section of water flow, the *maximum* air discharge is from geometrical considerations

$$Q_{aM} = [(A_t - A_w)/A_w]Q. \tag{2.30}$$

This has a maximum value because A_w is decreasing with increasing water discharge Q.

- If Δp is the subpressure behind the aeration pipe and V_o the approach velocity to the junction with the aeration conduit, i.e. $\mathbf{E} = [\Delta p/\rho g]/(V_o^2/2g)$ the Euler number of the aeration device, the air-water ratio $\beta = Q_a/Q$ depends on E, the Froude-number F and the relative tunnel length L_t/D_t with D_t as tunnel diameter.

• The aeration system should be designed for a *maximum air velocity* of $V_a = 50\,\text{ms}^{-1}$ in order that compressibility effects are negligible.

The aeration of air-water flow in vertical shafts without the 90° bend were investigated by Viparelli (1954). For unsubmerged flow in a smooth shaft of diameter D_s and length L_s, his results may be expressed with the Strickler roughness coefficient K_s as (Hager, 1994b)

$$\beta = \frac{1}{\pi}\left[\frac{Q}{\pi K_s D_s^{8/3}}[1 + (66D_s/L_s)]^{1/2}\right]^{-1/2} - 1. \qquad (2.31)$$

Figure 2.50 shows this relation and defines the shaft outlet geometry. Accordingly, β depends significantly on the relative discharge $q_s = Q/(\pi K_s D_s^{8/3})$ and slightly on the relative shaft length. The maximum air discharge Q_{aM} may be obtained from differentiation and reads

$$Q_{aM} = 0.011 K_s L_s^{1/3} D_s^{7/3} \qquad (2.32)$$

which occurs when $Q_{aM} = 0.41Q$.

Ervine and Himmo (1984) have considered the flow arrangement shown in Figure 2.51. A vertical shaft was connected by a sharp-edged mitre bend to a horizontal tunnel. The air pockets located in the tunnel are either transported in the tailwater, or they may rise back through to the shaft inlet.

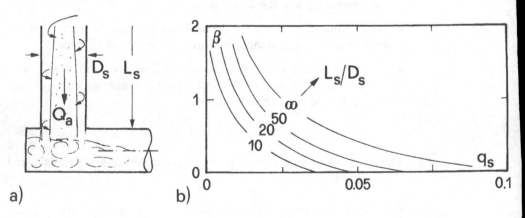

Figure 2.50 Air entrainment in vertical shaft (a) outlet geometry considered, (b) air-water ratio β as a function of relative discharge q_s for various relative shaft lengths L_s/D_s.

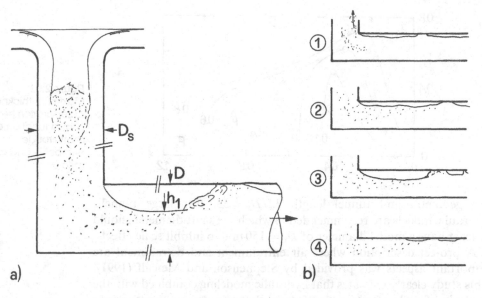

Figure 2.51 System shaft-tunnel (a) definition of geometry, (b) types of flow in tunnel spillway

According to energy considerations, the contracted depth h_1 downstream of the sharp-edged mitre bend of cross-section A_1 and Froude number $\mathbf{F}_s = V_s/(gD_s)^{1/2}$ is

$$h_1/D = 1 - \frac{1}{2}\mathbf{F}_s^2\left[\left(\frac{\pi D_s^2}{4A_1}\right)^2 - 1\right]. \tag{2.33}$$

If $\mathbf{F}_f = (Q_a + Q)/[(\pi D_s^2/4)(gD_s)^{1/2}]$ is the total discharge Froude number, various flow types may be identified (Figure 2.51(b)):

1. for $\mathbf{F}_f < 0.3$ air moves back to the shaft, and long air pockets develop in the tunnel;

2. for $0.25 < \mathbf{F}_f < 0.5$ and $\beta < 0.3$, i.e. relatively small air entrainment, no air backflow occurs and air pockets increase in length;

3. for $\mathbf{F}_f > 0.3$ and $\beta > 0.7(\mathbf{F}_f - 0.3)$ the flow in the tunnel may get supercritical with a transition to pressurized flow; and

4. for $\beta < 0.7(\mathbf{F}_f - 0.3)$ separation from the inner bend is inhibited, and air bubbles move along the vertex of the tunnel.

The *maximum separation height* $T_1 = 1 - h_1/D$ varies with the approach Froude number $\mathbf{F}_f$ and the air ratio β as shown in Figure 2.52. A maximum value establishes roughly for $\mathbf{F}_f = 0.3$. Ervine and Himmo point at the particular character of their results with regard to

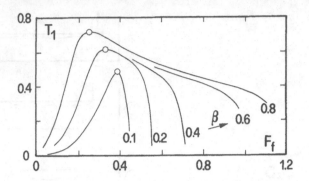

Figure 2.52
Maximum thickness
of separation height
T_1 as a function of
shaft Froude
number F_f and air
ratio β.

the geometry and tunnel length $L_s/D_s = 53$. For large tunnels, hydraulic models are recommended in which the shaft diameter should at least have a model diameter of $D_s = 150$ mm to inhibit scale effects.

A project description where air entrainment and detrainment are important aspects was provided by Stephenson and Metcalf (1991). This study clearly indicates that hydraulic modeling combined with the analysis of scale effects are currently the only design approach. In general, the *Froude similarity law* governs aeration processes in free surface flows. A contribution to the scaling of the tunnel flow downstream of a 90° bend was presented by Mussalli (1978). Again, a particular case was investigated and a generalization to the standard design is impossible.

2.5 SIPHON

2.5.1 Description of siphons

A siphon is a ducted overflow structure with either free surface or pressurized flow. Siphons are used as *spillway* either in parallel or in addition to other flood openings, or *intake* of small power plants, such as described by Xian-Huan (1989). In the following, the latter are not further considered.

Spillway siphons can either be a saddle siphon or a shaft siphon (Figure 2.53). Under increasing discharge both behave hydraulically like a weir. At a certain discharge priming occurs and the flow is pressurized for larger discharges. The transition between free surface and pressurized flow depends mainly on the aeration and deaeration of the siphon crest.

A siphon looks like a covered round-crested weir. Its advantages compared to the weir are the *priming action*, and a much increased discharge capacity for low heads. For larger heads, the weir has a larger capacity reserve in discharge, however (Figure 2.54). Therefore, siphons are suited at locations where the head is relatively small and regulation of discharge is not available. The disadvantages of a siphon are:

- causing of tailwater surges at rapid priming, as for sudden opening of a gate,
- cavitation damage by low pressure, and
- clogging by ice or wood.

The first two disadvantages can be removed by constructional means.

The *saddle siphon* is commonly used and is considered in the following. Siphons were incorporated in several regions into dams and weirs. There are few modern designs and there is a dearth of current literature. Detailed descriptions of siphons may be found, e.g. by Govinda Rao (1956), Press (1959), Samarin, et al. (1960) and Preissler and Bollrich (1985). The rehabilitation of a siphon with a relatively large head of 16 m is described by Bollrich (1994).

According to Head (1975) a distinction is made between the *black-water siphon* which can only operate on an on-off basis (i.e. full-bore, or no discharge at all) and the *air-regulated siphon* which automatically adjusts its discharge over the full range of flow to maintain a virtually *constant* headwater level. The latter inhibits flooding due to the abrupt

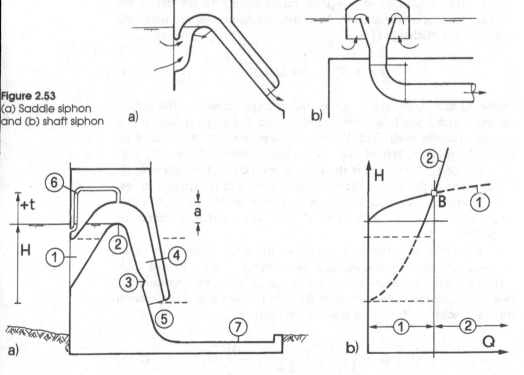

Figure 2.53
(a) Saddle siphon
and (b) shaft siphon

Figure 2.54 (a) Conventional siphon with ① intake, ② crest, ③ priming nose, ④ siphon barrel, ⑤ outlet, ⑥ aeration pipe, ⑦ stilling basin, (b) discharge head relation with ① weir regime, ② orifice regime. For equal width, a weir discharges less than a siphon in domain ①. At point B both structures have equal discharge

priming, and vibration problems. The characteristic of the air-regulated siphon is that it sucks in both water *and* air from where the white water originates. In the following both types of siphons are described.

2.5.2 Blackwater siphon

For this classic arrangement of a siphon, the transition between free-surface and pressurized flow occurs relatively abruptly. As the reservoir level increases over the siphon crest, the overflow starts by entraining air due to the large velocities. As soon as the inflow primes, the air is not substituted from the intake. If the downstream air access is also cut, such as by a jet deflector or priming nose (Figure 2.54) or by tailwater submergence, the siphon primes. An observer then sees *blackwater* at the siphon outlet.

The section of a saddle siphon is often rectangular, with a typical height to width ratio of 1:1.5 to 1:2.5. To reduce oscillations in the reservoir elevation and for an adequate hydraulic inlet shape, an aeration conduit can be added, with a typical cross-section 3 to 5% of the siphon crest section.

The *hydraulic design* of a siphon refers mainly to the maximum discharge, i.e. pressurized flow. According to the generalized Bernoulli equation the discharge Q is

$$Q = A_d V_d = ab\eta(2gH_0)^{1/2} \tag{2.34}$$

where A_d and V_d are cross-section and average velocity of the siphon outlet, a and b are height and width at the siphon crest, and η is a siphon efficiency coefficient. For $\eta = 1$, the siphon flow would have no losses. In practice, typical values of η are between 0.7 and 0.9, with $\eta = (1 + \Sigma\xi_i)^{-1/2}$ where ξ_i is the ith loss coefficient. The sum over all losses involves the intake, the bend, the outlet and the friction losses. For the siphon according to Figure 2.55 for example, Bollrich (1994) determined $\eta = 0.86$ for a head of $H_o = 16\,\mathrm{m}$ and a discharge of $Q=51.7\mathrm{m}^3\,\mathrm{s}^{-1}$.

In a second step, the subpressure at the crest section has to be checked against *cavitation damage*. By analogy to the standard spillway, the minimum pressure occurs at the siphon crest. Assuming a flow with concentric streamlines in the crest region yields an expression for the discharge Q_c with incipient cavitation as

$$Q_c = br_i \ln\left(\frac{r_a}{r_i}\right)\left[\frac{2g(h_A - h_c + t)}{1 + \Sigma\xi_i}\right]^{1/2}. \tag{2.35}$$

Here, r_i and r_a are inner and outer crest radii, respectively, $h_A[\mathrm{m}] = 10.3 - (A/900)$ is the atmospheric pressure head at elevation

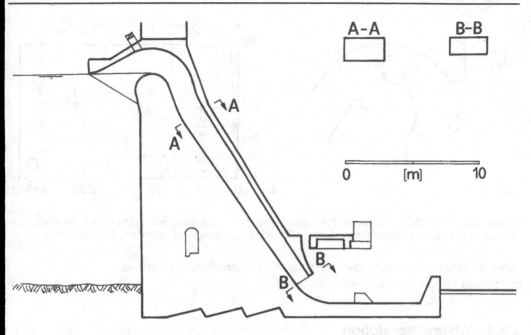

Figure 2.55 Siphon spillway of Burgkhammer dam (Germany) according to Bollrich (1994). The pairs of siphons have equal crest but different inlet nose elevations

A above sea level, h_c is the pressure head at incipient cavitation, t is the reservoir elevation above the siphon crest and $\Sigma \xi_i$ is the sum of all head loss coefficients, as discussed previously. In general, a conservative value of $h_c = 2$ m is assumed. If the maximum discharge Q_M is larger than the cavitation discharge Q_c, then the siphon has to be modified until the condition $Q_M < Q_c$ is satisfied.

For heads larger than, say 8 m on the siphon, the barrel has to be *contracted* from the crest section towards the outlet. This improves the pressure conditions in the crest section. To shorten the barrel, the outlet can also be located above the tailwater level. The flow is discharged over a flip bucket into a plunge pool, or over a spillway into a stilling basin (Chapter 5).

To reduce the risk of tailwater flooding during siphon priming, a number of *siphons in parallel* can be arranged. The siphon crests are located at increasing elevations with a successive priming. Figure 2.56 refers to an example of six individual siphons two of which have identical priming conditions. The difference of head to the next pair of siphons is 0.1 m. The resulting discharge-head curve is shown in Figure 2.56. Note the differences between the curves for increasing, and decreasing discharge as is typical for siphons.

Under adverse discharge conditions, siphons may vibrate. A discussion is given by Govinda Rao (1956). Other types of overflow struc-

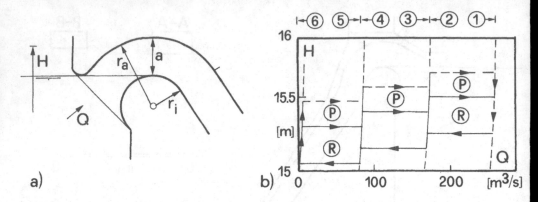

Figure 2.56 (a) Definition of siphon crest geometry, (b) siphon operation cycle with *P* priming and *R* reduction of flow for Burgkhammer siphon with a total discharge of $250\,\text{m}^3\,\text{s}^{-1}$ (Bollrich, 1994)

tures such as gates may also be subject to vibrations. A general discussion on the topic follows in section 6.3.

2.5.3   Whitewater siphon

Figure 2.57 shows a typical structure, with an upward sloping converging inlet, a covered crest of almost prismatic geometry and a downward sloping outlet. The *priming nose* seals the siphon pipe against the atmosphere. The head on the whitewater siphon may vary between 0.5 m and 2 m and the capacity is up to $4\,\text{m}^3/\text{sm}$. The manometer tapping is not essential but provides a useful means of monitoring performance.

The *discharge-head relation* was described by Head (1975) in three ranges, namely the weir flow, the air-regulated flow and the drowned flow. The hydraulic features are (Figure 2.58):

① *weir flow* behaves like the overflow over a cylindrical crest under atmospheric air pressure in the pipe;

② *air-regulated flow* starts when the *barrel* is sealed from the atmosphere due to the jet deflection. The evacuation of air continues and a partial vacuum is created that increases the discharge without a significant headwater level rise. Thus, air can be continuously sucked into the siphon to compensate for the extraction at the siphon outlet. A rise in the headwater yields a reduced air flow, and a fall leads to an increased air flow and thus a reduced discharge. Accordingly, the whitewater siphon is *self-regulating*; and

③ *drowned flow*, including the transition from the air partialized range with an air-water mixture in the barrel, to full-pipe flow. The discharge then may be computed from the generalized Bernoulli equa-

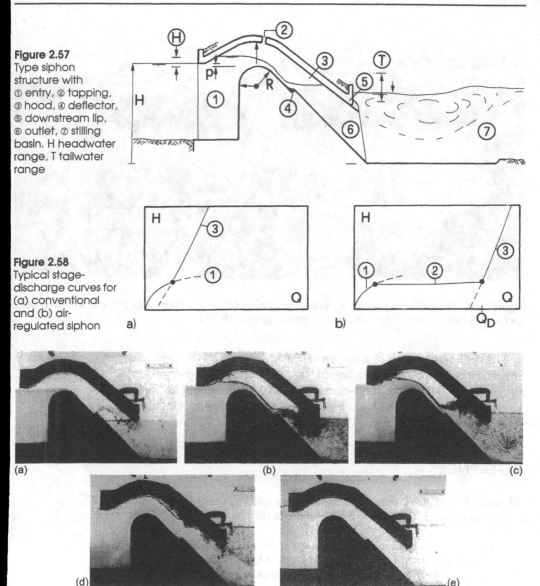

Figure 2.57
Type siphon
structure with
① entry, ② tapping,
③ hood, ④ deflector,
⑤ downstream lip,
⑥ outlet, ⑦ stilling
basin. H headwater
range, T tailwater
range

Figure 2.58
Typical stage-
discharge curves for
(a) conventional
and (b) air-
regulated siphon

Figure 2.59 Flow phases in air-regulated siphon (a) weir flow, (b) deflected nappe, (c) drowned
nappe, (d) air-partialized flow, (e) drowned flow (Head 1975)

tion involving the difference level between headwater and tailwater
elevations as given in 2.5.2. The air regulation has ceased and this
range seldom occurs, e.g. for an exceptional flood. The design
should definitely account for the transition discharge between air-
regulation and drowning flow.

Figure 2.59 shows the five stages of an air-regulated siphon.

Figure 2.60 Downstream view of Barikese dam, Ghana (Communication of J.D. Williams on a Self-regulating river syphon, *Journal Institution Water Engineers* **25**: 213)

The crest positioning of a siphon spillway depends on the particular site criteria. Once a siphon begins to operate in the air-regulating range, it will flatten the stage-discharge curve. Both Head (1975) and Ervine and Oliver (1980) have provided detailed design guidelines. Figure 2.60 refers to a typical siphon structure.

REFERENCES

Abecasis, F.M. (1970). Discussion to Designing spillway crests for high-head operation, by J.J. Cassidy. *Journal Hydraulics Division* ASCE **96**(12): 2654–2658.

Bollrich, G. (1994). Hydraulic investigations of the high-head siphon spillway of Burgkhammer. *18 ICOLD Congress* Durban **Q71**(R2): 11–20.

Bradley, J.N. (1956). Shaft spillways – prototype behavior. *Trans. ASCE* **121**: 312–344.

Bretschneider, H. (1980). Kreisförmige Fallschächte für die Hochwasserentlastung bei Talsperren (Morning Glory shaft for spillways on dams). *Wasserwirtschaft* **70**(3): 88–93 (in German).

Bretschneider, H., and Krause, D. (1965). Die Modellversuche für die Hochwasserentlastungsanlage der Innerste-Talsperre im Harz (The scale experiments for the spillway of the Innerste dam in the Harz). *Mitteilung* **62**. Institut für Wasserbau und Wasserwirtschaft, Technische Universität: Berlin (in German).

Chow, V.T. (1959). *Open channel hydraulics.* McGraw-Hill, New York.

Creager, W.P., Justin, J.D., Hinds, J. (1945). *Engineering for dams.* 2. John Wiley, Chapman & Hall, London.

Ervine, D.A. and Himmo, S.K. (1984). Modelling the behaviour of air pockets in closed conduit hydraulic systems. *Scale Effects in Modelling Hydraulic Structures* Esslingen 4(15): 1–12.

Ervine, D.A. and Olivier, G.C.S. (1980). The full-scale behaviour of air-regulated siphon spillways. *Proc. Institution Civil Engineers* 69(2): 687–706.

Gardel, A. (1949). Les évacuateurs de crues en déversoirs circulaires (The spillways for circular overfalls). *Bulletin Technique de la Suisse Romande* 75(27): 341–349 (in French).

Govinda Rao, N.S. (1956). Design of siphons. *Publication* 59. Central Board of Irrigation and Power: New Delhi.

Hager, W.H. (1987). Continuous crest profile for standard spillway. *Journal Hydraulic Engineering* 113(11): 1453–1457.

Hager, W.H. (1990). Vom Schachtüberfall zum Schachtwehr (From shaft weir to shaft spillway). *Wasserwirtschaft* 80(4): 182–188 (in German).

Hager, W.H. (1991a). Experiments on standard spillway flow. *Proc. Institution Civil Engineers* 91: 399–416.

Hager, W.H. (1991b). Flow features over modified standard spillway. *24 IAHR Congress* Madrid D: 243–250.

Hager, W.H. (1994a). *Abwasserhydraulik* (Sewer hydraulics). Springer, Heidelberg (in German).

Hager, W.H. (1994b). Wasser-Luftgemische in Vertikalrohren (Air-water mixtures in vertical pipes). *gwf/Wasser-Abwasser* 135(7): 391–397 (in German).

Hager, W.H. and Bremen, R. (1988). Plan gate on standard spillway. *Journal Hydraulic Engineering* 114(11): 1390–1397.

Hartung, F. (1973). Gates in spillways of large dams. *11th ICOLD Congress* Madrid: Q41(R72): 1361–1374.

Head, C.R. (1975). Low-head air-regulated siphons. *Journal Hydraulics Division ASCE* 101(HY3): 329–345; 102 (HY1): 102–105; 102 (HY3): 422–425; 102 (HY10): 1597.

ICOLD, (1987): Spillways for dams. *Bulletin* 58. International Commission for Large Dams, Paris.

Indlekofer, H. (1976). Zum hydraulischen Einfluss von Pfeileraufbauten bei Überfall-Entlastungsanlagen (The hydraulic effect of piers on overfall dams). *Mitteilung* 13. Institut Wasserbau und Wasserwirtschaft, Rheinisch-Westfälische Technische Hochschule, Aachen (in German).

Indlekofer, H. (1978). Zur Berechnung der Überfallprofile von kelchförmigen Hochwasserentlastungen (The computation of crest profiles of cup-shaped dam spillways). *Bautechnik* 55(11): 368–371 (in German).

Lemos, F. de Oliveira (1981). Criterios para o dimensionamento hydraulico de barragens descarregadioras (Criteria for the hydraulic design of gated spillways). *Memoria* 556. Laboratorio Nacional de Engenharia Civil LNEC, Lisbon (in Portuguese).

Mussalli, Y.G. (1978). Size determination of partly full conduits. *Journal Hydraulics Division ASCE* 104(HY7): 959–974; 105(HY8): 1039–1041.

Novak, P. and Cabelka, J. (1981). *Models in hydraulic engineering*. Pitman, Boston and London.

Peterka, A.J. (1956). Shaft spillways – Performance tests on prototype and model. *Trans. ASCE* 121: 385–409.

Preissler, G. and Bollrich, G. (1985). *Technische Hydromechanik* (Technical hydromechanics) 1. VEB Verlag für Bauwesen, Berlin (in German).

Press, H. (1959). *Wehre* (Weirs). Wilh. Ernst & Sohn, Berlin (in German).

Reinauer, R. and Hager, W.H. (1994). Supercritical flow behind chute piers. *Journal Hydraulic Engineering* **120**(11): 1292–1308.

Rhone, T.J. (1959). Problems concerning use of low head radial gates. *Journal Hydraulics Division* ASCE **85**(HY2): 35–65; **85**(HY7): 151–154; **85**(HY9): 113–117; **86**(HY3): 31–36.

Samarin, E.A., Popow, K.W., Fandejew, W.W. (1960). *Wasserbau* (Translation from Russian). VEB Verlag für Bauwesen: Berlin (in German).

Sinniger, R.O. and Hager, W.H. (1989). *Constructions hydrauliques* (Hydraulic structures). Presses Polytechniques Fédérales, Lausanne (in French).

Smith, W. (1966). Bellmouth spillway and stilling basin – hydraulic model tests for Lower Shing Mun reservoir. *Civil Engineering and Public Works Review* **61**(3): 302–305; **61**(4): 499–501.

Stephenson, D. and Metcalf, J.R. (1991). Model studies of air entrainment in the Muela drop shaft. *Proc. Institution Civil Engineers* **91**(2): 417–434.

US Corps of Engineers (1970). *Hydraulic design criteria.* Army Waterways Experiment Station, Vicksburg, MI.

USBR (1938). Model studies of spillways. *Bulletin* **1.** Boulder Canyon Projects, Final Reports. Part VI – Hydraulic Investigations. US Department of Interior, Bureau of Reclamation, Denver.

Viparelli, M. (1954). Trasporto di aria da parte de correnti idriche in condotti chiusi (Air transport as a part of a hydraulic current in a closed conduit). *L'Energia Elettrica* **31**(11): 813–826 (in Italian).

Wagner, W.E. (1956). Determination of pressure-controlled profiles. *Trans. ASCE* **121**: 345–384.

Webster, M.J. (1959). Spillway design for Pacific Northwest projects. *Journal Hydraulics Division* ASCE **85**(HY8): 63–85.

Xian-Huan, W. (1989). Siphon intakes for small hydro plants in China. *Water Power and Dam Construction* **41**(8): 44–53.

Seujet free surface diversion channel, Geneva (Courtesy Division des Ponts et des Eaux, Genève)

3

Diversion Structures

3.1 INTRODUCTION

A dam is a major structure closing the lower portion of a valley. Both during construction and later in service, the dam should not completely block the valley. For example, floods may arrive during construction and are then particularly dangerous, because the structure is not finalized and thus prone to damage. The design discharge for dam diversion has been determined in Chapter 1. Although the structures under discussion are often provisional during construction, they should be designed correctly. Their failure may have catastrophic consequences. A substitute waterway, the so-called *dam diversion,* must be ready in order that the river can bypass the dam site and the floods cause no harm to the area under construction (Figure 3.1).

In 3.2 and 3.3, two types of diversions are presented: a *diversion tunnel* as a partially filled spillway, and a diversion channel as a local river constriction. A main concern in tunnel hydraulics is the transition from free surface to pressurized flow, particularly in regard to the design discharge and the definition of the tunnel roughness. Also, moving hydraulic jumps may produce unstable tunnel flows.

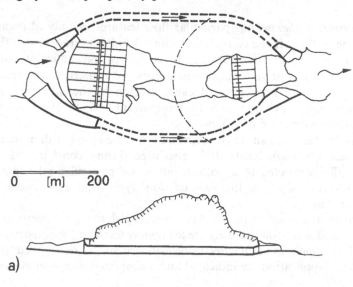

Figure 3.1
Dam diversion during construction for Cabora Bassa dam on Zambesi river (Africa).
(a) Plan and section of right diversion tunnel, (b) final dam (*ICOLD* **Q.50**, R.41)

0   [m]   200

a)

b)

Figure 3.1
(Continued)

A *diversion channel* has particular flow features, mainly as regards
the type of flow. If the constriction rate is too large, transitional flow
may occur, and the structure can be compared hydraulically with a
Venturi flume. For subcritical flow across the constriction, standing
waves may be a nuisance and cause erosion in the tailwater. The
upstream head-discharge curve is an important basis for the economic
and safe construction of a dam.

A *culvert* is a transition or diversion structure across a dam mainly
for small upstream heads. It is also used during construction and
eventually connected to a bottom outlet. Of particular relevance in
the culvert design are the various flow types that may occur, as
described in 3.4.

Diversion tunnels bypass a river around a dam site, culverts cross
the site and diversion channels are led merely through the construction
site. The principles of all three transition structures are outlined, their
ranges of application are indicated and a simplified design guideline is

given. There are a number of additional points to be considered for a particular structure, mainly with regard to erosion and sedimentation, and the outlet pattern. These items are site-specific, however, and a generalized design can hardly be given. A number of relevant references is listed for further reading.

3.2 DIVERSION TUNNEL

3.2.1 Introduction

Figure 3.2 shows a photograph of a typical intake of a diversion tunnel. The dam's construction site is protected by two *cofferdams*: The upstream dam guides the water away from the valley into the diversion tunnel and the downstream cofferdam returns the flow back to the river. For larger design floods, several diversion tunnels may be arranged, such as in Figure 3.2. Diversion tunnels can also run around both sides of a valley. Normally, diversion tunnels have a *free-surface flow* to divert floating matter also. The cross-sectional geometry and the layout are designed for their final purpose: Either, the tunnel can be closed after construction, or it can be connected to a bottom outlet, a withdrawal structure or a water intake.

Figure 3.2 Entry to diversion tunnel of Karakaya dam (Turkey) during construction (*Hydraulic Engineering Works* Italstrade: Milan 1990)

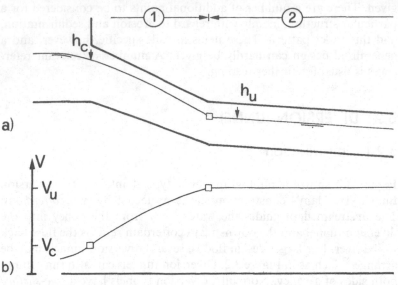

Figure 3.3
Inlet of diversion
tunnel (a) section,
(b) streamwise
increase of average
velocity. ①
Acceleration reach,
② equilibrium reach.

The layout of the diversion tunnel depends on the available slope. For small slopes, there is practically no margin, whereas for larger slopes, the tunnel has an acceleration reach followed by an equilibrium reach (Figure 3.3). The flow is accelerated from the intake to the critical velocity V_c at the grade break to the uniform velocity V_N, along the equilibrium reach. If the acceleration reach is practically a spillway, then the slope can be relatively large. For excavated tunnels, the maximum slope is typically 10% due to constructional reasons. The slope in the equilibrium reach should satisfy the following conditions:

- reduction of the *cross-section* for economic reasons,

- generation of *stable supercritical flow*, i.e. with a Froude number larger than 1.5 to 2 to inhibit undular flow,

- allow for *transport of sediment*, if it is not retained upstream from the intake, and

- prevention of *tunnel abrasion* (Vischer, et al., 1997).

The *diversion tunnel* consists of three major portions:

- the inlet,

- the tunnel, and

- the outlet.

Each of these three portions has hydraulic particularities that are summarized below.

3.2.2 Inlet to diversion tunnel

The purpose of the *inlet structure* is to accelerate the flow to the tunnel velocity, to yield a smooth transition from the river to the tunnel flow and to provide sufficient air for atmospheric tunnel pressure. Because the cross-sectional shape of the river is nearly trapezoidal and the tunnel is either of horseshoe or of circular shape, a complicated transition shape is needed for acceptable flow conditions. Hinds (1928) proposed a warped wall inlet (Figure 3.4(a)). For small and medium discharges a simpler structure was proposed by Smith (1967), the length L of which is only $1.25(B-b)$, where B and b are the upstream bottom width, and the tunnel width, respectively (Figure 3.4(b)). The entry radius is $R_e = 1.65(B - b)$. The energy loss ΔH_e across the entry was expressed in terms of the tailwater velocity $V_t = Q/(bh_t)$ and the width ratio $\beta_e = b/B$

$$\Delta H_e = 0.06(1 - \beta_e)\frac{V_t^2}{2g}. \tag{3.1}$$

In order to prevent a standing wave pattern, the Froude number $F_t = V_t/(gh_t)^{1/2}$ based on the tailwater depth should not exceed 0.67.

For a *transition* from sub- to supercritical flow at the inlet, the location of critical flow can be determined with the theorem of Jaeger (Jaeger, 1949). The location of critical flow is where the energy head H has a minimum (Figure 3.5). The free surface profile can be determined with conventional backwater curves, starting at the critical point with the critical depth h_c, moving in the upstream direction for subcritical, and in the downstream direction for supercritical flow.

3.2.3 Tunnel flow

The free surface flow in a tunnel may be computed as an open channel and involves *backwater curves*. Figure 3.6 shows a definition sketch

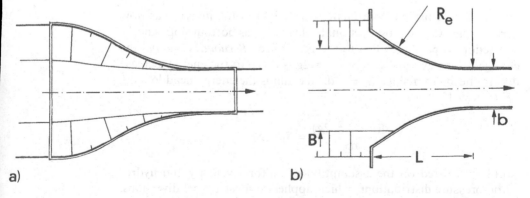

a)   b)

Figure 3.4 Inlet structures according to (a) Hinds (1928) and (b) Smith (1967)

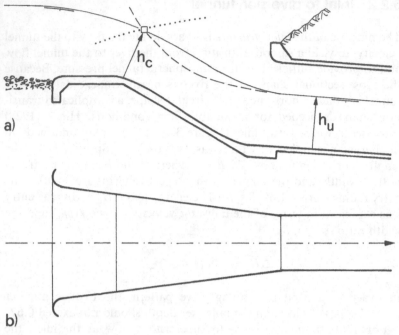

Figure 3.5
Transition from sub-
to supercritical flow
in a diversion inlet
with an increase of
bottom slope and a
decrease of width.
(a) section with
(– – –) uniform flow
depth profile, (…)
critical flow depth
profile, (b) plan.

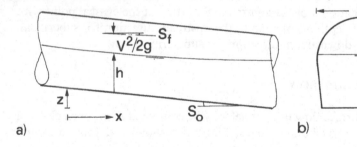

Figure 3.6
Backwater curves in
diversion tunnel
(a) longitudinal
section, (b)
transverse section

with x as longitudinal coordinate, z as height of the invert, h as flow
depth, V as average cross-sectional velocity, S_o as bottom slope and S_f
as friction slope. According to the generalized *Bernoulli equation*, the
change of energy head $H = h + V^2/2g$ is equal to the energy increase
due to the bottom slope $S_o = -dz/dx$ minus the energy head loss S_f,
i.e. (Chow 1959)

$$\frac{d(h + V^2/2g)}{dx} = S_o - S_f. \tag{3.2}$$

Eq.(3.2) is based on the assumption of uniform velocity and hydrostatic pressure distributions, which applies to most tunnel diversions.
The *Froude number*

$$\mathbf{F} = V/c \tag{3.3}$$

is defined as the ratio of average velocity V to the propagation velocity c of a shallow water wave. The latter is equal to the square root of the cross-sectional area A divided by the free surface width $B_s = \partial A/\partial h$ times the gravitational acceleration g, i.e.

$$c = [gA/(\partial A/\partial h)]^{1/2}. \tag{3.4}$$

Equating the latter two relations yields the equation of backwater curves in the *prismatic* channel

$$\frac{\mathrm{d}h}{\mathrm{d}x} = \frac{S_o - S_f}{1 - \mathbf{F}^2}. \tag{3.5}$$

The backwater curves are dominated by two particular cases:

1. $S_o = S_f$, i.e. the *uniform flow* condition, for which the flow depth h does not change along the channel because $\mathrm{d}h/\mathrm{d}x = 0$, and

2. $\mathbf{F}^2 = 1$, i.e. the *critical flow* condition, for which the surface profile is theoretically vertical because $\mathrm{d}x/\mathrm{d}h = 0$.

The uniform flow corresponds to the equilibrium of tractive and resisting forces, comparable to a condition of static flow. It is governed by the bed slope of the channel and the fluid viscosity, the roughness pattern of the channel walls, the velocity and the channel geometry. The uniform flow is described by the resistance law as given below. It may also be identified by the *uniform flow depth* h_N.

The critical flow corresponds to the condition of minimum energy head, and the *transition* between the subcritical and the supercritical flow. The critical flow is also characterized by the *critical flow depth* h_c. For subcritical flow, the velocity V is smaller than the propagation velocity c, or the flow depth h is larger than the critical flow depth h_c and perturbations propagate both in the upstream and downstream directions. Subcritical flow may be described as smooth, slow and nearly one-dimensional for which the backwater curves are a good approximation.

For supercritical flow, in contrast, the propagation velocity c is smaller than the tunnel velocity V, or $h < h_c$ in terms of flow depths, and perturbations propagate only in the flow direction. Such flows can be characterised as rough, fast and two-dimensional, because of the oblique surface pattern behind any perturbation of flow. Figure 3.7 shows a curved tunnel flow with so-called *shockwaves* along which all perturbations are propagated. The shockwaves tend to flow concentrations and a supercritical flow can be improved by making the surface pattern uniform (chapter 4).

The *friction slope* S_f for continuous flows can be computed according to the resistance laws of hydraulics. The classic resistance equation according to Darcy-Weisbach reads

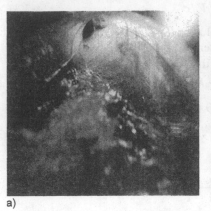
a)

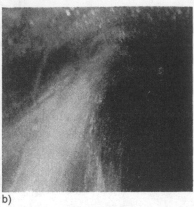
b)

Figure 3.7 Flow in a curved tunnel spillway with the formation of shockwaves. Views from (a) upstream, (b) downstream

$$S_f = \frac{f}{D_h} \frac{V^2}{2g} \tag{3.6}$$

where f is the friction factor, D_h the hydraulic diameter defined as four times the hydraulic radius $R_h = A/P_h$ with P_h as wetted perimeter, and $V^2/2g$ the velocity head. The *friction factor f* depends on the:

- resistance characteristics of the fluid, i.e. the kinematic viscosity, and
- resistance characteristics of the boundary material.

According to the concept of *equivalent roughness*, a uniform sand roughness height k_s [m] causes the resistance effect identical to an arbitrary roughness pattern. The scaling length of the roughness height is the hydraulic diameter D_h, and $\epsilon = k_s/D_h$ is the relative roughness. Colebrook and White (1937) proposed the *universal friction law*

$$f^{-1/2} = -2\log\left[\frac{\epsilon}{3.7} + \frac{2.51}{\mathbf{R}f^{1/2}}\right] \tag{3.7}$$

where $\mathbf{R} = VD_h/\nu$ is the *Reynolds number*. A flow is turbulent if $\mathbf{R} > 2300$. In civil engineering applications, flows are always turbulent. For $[\epsilon/3.7] \ll [2.51/\mathbf{R}f^{1/2}]$ wall roughness is insignificant and the flow is said to be in the hydraulic *smooth* regime. In contrast, the hydraulic *rough* regime occurs provided $[\epsilon/3.7] \gg [2.51/\mathbf{R}f^{1/2}]$. If both wall roughness and viscosity effects are significant, the flow is in the transition regime.

Because the exact definition of the equivalent roughness height k_s is often difficult, and due to the small effect of viscosity for flows in hydraulic structures, the usual approach is to assume flows as *turbulent rough*. Then, f depends exclusively on the relative roughness as

$$f^{-1/2} = -2\log[\epsilon/3.7]. \tag{3.8}$$

The relevant roughness domain is approximately between $5.10^4 < \epsilon < 5.10^{-2}$. For usual Reynolds numbers $\mathbf{R}$ from 10^5 to 10^7 Eq. (3.8) may be approximated with a power function as

$$V = KS_f^{1/2}R_h^{2/3}. \tag{3.9}$$

This is the *GMS formula* deduced by the Frenchman *Gaspard-Philibert Gauckler* (1826–1905), redeveloped by the Irishman *Robert Manning* (1816–1897) and consecutively tested with additional data by the Swiss *Albert Strickler* (1887–1963). The formula is simple in application, and thus popular. Either of the quantities may explicitly be computed, and the Strickler roughness value K, or the Manning friction factor $1/n = K$ are well-known dimensional quantities to the experienced hydraulic engineer. Table 3.1 provides a summary of n-values, and more detail is given, e.g. by Chow (1959) or Schröder (1990). The actual value of the roughness coefficient is a matter of guess work, finally, and one should carefully select the appropriate roughness coefficient based on site conditions, experience, and significance of the structure.

Gravel bed rivers that are typical for diversion channels are characterized by the gravel diameter d_n, where n is the percentage of grains by weight that are smaller than indicated by the value n. For bottom slopes $4 \cdot 10^{-3} < S_o < 2.5 \cdot 10^{-2}$ and hydraulic radii $10^{-1}\text{m} < R_h < 10^1\text{m}$ Strickler (1923) proposed the empirical relation

$$K = 1/n = 21.1d_{50}^{-1/6}. \tag{3.10}$$

Increasing the grain diameter thus gives an increase in roughness. Meyer-Peter and Müller (1948) related their considerations to the characteristic grain diameter d_{90} instead of d_{50}, i.e. to almost the largest sediment size on the riverbed, and proposed in dimensionless form

$$\frac{(1/n)d_{90}^{-1/6}}{g^{1/2}} = 8.2 \tag{3.11}$$

Neill, who discussed Bray and Davar (1987) proposed the simple dimensional relation

$$1/n = 10S_o^{-1/6} \tag{3.12}$$

and recommended this check equation for all calculations. Clearly, the grain diameter varies with the river bottom slope and so does the roughness parameter.

Jarrett (1984) has recommended for streams of average slope larger than 0.2% the formula

$$V = 3.81S_o^{0.21}R_h^{0.83}. \tag{3.13}$$

Table 3.1 Roughness coefficient n according to Manning, and $K = 1/n$ $[m^{\frac{1}{3}}s^{-1}]$ according to Strickler (adapted from Chow, 1959)

Type			Minimum	Normal	Maximum
A CLOSED CONDUIT FLOWING PARTLY FULL					
Metal	Brass, smooth		0.009	0.010	0.013
	Steel,	lockbar and welded	0.010	0.012	0.014
		reveted and spiral	0.013	0.016	0.017
	Cast iron, coated		0.010	0.013	0.014
	Corrugated metal, subdrain		0.017	0.019	0.021
Nonmetal	Cement,	neat	0.010	0.011	0.013
		mortar	0.011	0.013	0.015
	Concrete,	straight and clean	0.010	0.011	0.013
		with bends, some debris	0.011	0.013	0.014
		unfinished, steel form	0.012	0.013	0.014
		unfinished, rough	0.015	0.017	0.020
B LINED OR BUILT-UP CHANNELS					
Metal	smooth surface		0.011	0.012	0.014
	corrugated		0.021	0.025	0.030
Nonmetal	Cement, smooth		0.010	0.011	0.013
Concrete	Trowel finish		0.011	0.013	0.015
	Float finish		0.013	0.015	0.016
	unfinished		0.014	0.017	0.020
	Gunite, good section		0.016	0.019	0.023
	good excavated rock		0.017	0.020	–
	irregular rock		0.022	0.027	–
	vegetal lining		0.030	.....	0.500
C EXCAVATED OR DREDGED					
Earth	straight and uniform and clean		0.016	0.018	0.020
	winding and sluggish	no vegetation	0.023	0.025	0.030
		weeds	0.030	0.035	0.040
		stony bottom	0.025	0.035	0.040
	rock cuts	smooth and uniform	0.025	0.035	0.040
		jagged and irregular	0.035	0.040	0.050
	not maintained	dense weeds	0.050	0.080	0.120
		dense bush, high stage	0.080	0.100	0.140
D NATURAL STREAMS					
Minor streams	clean and straight		0.025	0.030	0.033
	clean, winding, with pools		0.033	0.040	0.045
	clean, but with cobbles		0.045	0.050	0.060
	weedy, deep pools, underbrush		0.075	0.100	0.150
	mountain stream	gravel	0.030	0.040	0.050
		boulders	0.040	0.050	0.070
Flood plains	no brush, short grass		0.025	0.030	0.035
	cultivated areas, no crop		0.020	0.030	0.040
	brush	scattered	0.035	0.050	0.070
		dense	0.070	0.100	0.160
Major streams	regular		0.025	.....	0.060
	irregular		0.035	.....	0.100

It was calibrated for data collected in the Colorado river basin for gravel rivers up to 4% slope and for hydraulic radii smaller than 2.1 m.

Rock tunnels as occur in diversion tunnels have received particular attention, because of:

- their largeness for which the universal friction law was not tested,

- their particular surface patterns, and

- the effect of deposits and organic growths.

The contributions of Colebrook (1958), ASCE Task Force (1965), and Barr (1973) revealed that the roughness of large tunnels is sometimes considerably larger than anticipated. Algal growth on the concrete lining surfaces increased the flow resistance and caused it to vary seasonally. The design of large tunnels should thus at least include an assessment of the minimum and maximum friction coefficients. Also, an initially smooth surface may become rough due to sediment deposits or abrasion.

3.2.4 Outlet structure

The outlet of a diversion tunnel is either connected to the river directly, or a stilling basin is provided to dissipate excess kinetic energy (Figure 3.8). The later structure is needed either when the tunnel flow is supercritical or the elevation of the tunnel outlet relative to the river bottom is larger than needed for subcritical flow. The design of the junction structure and the river depends on site conditions, and details cannot be given in general. There are two aspects of importance, however:

Figure 3.8
Outlet of by-pass for Adda river (Valtellina, Italy) due to large rock slide in 1987, pipe diameters 4.2 and 6.0 m (*Italstrade* Milan, 1990)

1. If possible, the tailwater level should be so low that the tunnel flow is *not submerged*. The tunnel may seal otherwise and yield complex air-water flows associated with underpressure.

2. If the approach Froude number to the tailwater stilling basin is small problems with the stability of dissipation may arise, e.g. weak jumps, asymmetric flow, blow out and tailwater waves.

If the diversion tunnel is subsequently used as a tunnel spillway, such as for a Morning Glory spillway, provision for the final design should be made. A trajectory spillway (Chapter 5) is a common design for locations with space limitations.

3.3   RIVER DIVERSION

3.3.1   Effect of constriction

Instead of diverting a river in a tunnel or in a culvert, as previously discussed, one may also laterally constrict the river and start construction in the contracted river portion, provided site conditions allow this procedure. In a second stage, the already finished part can be given back to the river and the construction may proceed to the adjacent portion. Often, several stages are necessary. Figure 3.9 shows the diversion channel of Clyde dam. The channel was designed for a discharge of $1800\,m^3s^{-1}$ corresponding to the flood with a fifteen year return period. The freeboard is $1\,m$ below the crest of the upstream cofferdam. In November 1984 a flood with $2200\,m^3s^{-1}$ was diverted without problems (Vischer, 1987).

Figure 3.9
Clyde hydro-power station NZ. (a) river diversion to allow for construction,

a)

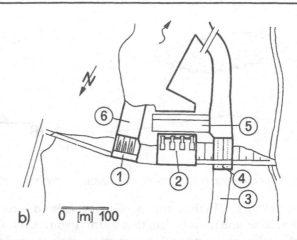

Figure 3.9
(b) plan of final
design with
① spillway,
② penstocks,
③ diversion channel,
④ diversion sluices,
⑤ powerhouse,
⑥ stilling basin.

b) 0 [m] 100

Figure 3.10 refers to Al-Massira multipurpose scheme in Morocco. The buttress dam has a height of 79 m and the design discharge of the spillway is 6000 m^3s^{-1}, and 1380m^3s^{-1} may be diverted with the bottom outlet. The diversion channel is located on the left river side and has a

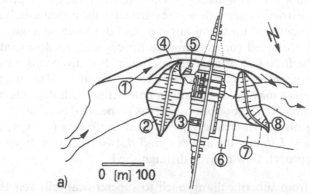

a) 0 [m] 100

Figure 3.10
Al-Massira dam,
Morocco. (a) Plan
with ① diversion
channel, ②
upstream
cofferdam, ③
intakes, ④ spillway
and bottom outlet,
⑤ diversion tunnel, ⑥
powerhouse, ⑦
tailrace channel, ⑧
downstream
cofferdam (b)
intake and dam
under construction

b)

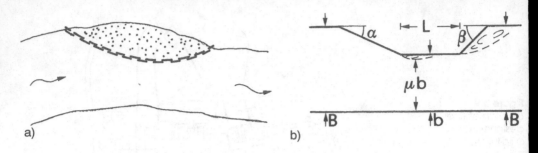

Figure 3.11 River diversion (a) typical configuration, (b) hydraulic abstraction

capacity of $2600 \text{m}^3\text{s}^{-1}$. In the centre portion it consists of a rectangular channel 18 m wide, and the total length is 400 m (Figure 3.10(a)). Due to sound rock quality, it was possible to excavate nearly vertical walls.

The effect of a *constriction* on the river flow may be significant, and a hydraulic design has to account for the submergence effect, in order that works under floodwater can proceed. Often, the river at the dam site can be approximated as a rectangular channel of approach width B and constricted width b (Figure 3.11). The transition from the approach river to the constriction depends on the particular site conditions, but a polygonal geometry of upstream angle α, and downstream angle β is quite general. Rounded corners improve the capacity, as described at the end of the following simplified approach. For diversions with a more complex geometry, scale modelling is recommended. The effect of a loose boundary may be important in connection with deposits and erosion of sediment. The bottom roughness is specified in 3.2.6.

Depending on the discharge Q, the width ratio $\psi = b/B \le 1$, the length ratio $\lambda = L/b$ and the angles α and β, two basic flow types for subcritical approach flow may be distinguished:

- transition from subcritical approach to supercritical tailwater flow for a strong constriction, and

- subcritical flow throughout for a weak constriction.

The second case is typical, because a strong constriction causes a significant submergence into the approach river, and scour problems in the tailwater. Both cases are discussed subsequently.

3.3.2 Transitional flow

For a transition from sub- to supercritical flow, *critical flow* is known to occur at the location of minimum flow width. For a polygonal constriction, the minimum width occurs slightly downstream from the leading edge, and the corresponding contraction coefficient is μ (Figure 3.11(b)). Based on the momentum equation, Hager and Dupraz (1985) found that

$$\mu = \hat{\mu} + \frac{1}{3}(1 - \hat{\mu})(\psi + 2\psi^4), \tag{3.14}$$

$$\hat{\mu} = (1 + \lambda)^{-1}\left[\frac{1 + \sin(\alpha/2)}{(1 + \sin(\alpha/2))^2} + \lambda\left(1 - \frac{3}{8}\sin^{0.8}(\alpha/2)\right)\right]. \tag{3.15}$$

Clearly, the tailwater angle β has no effect on the contraction coefficient μ. Figure 3.12(a) shows $\mu(\psi)$ for various angles α and $\lambda = 1$. For $\psi = 1$ or $\alpha = 0$ one has evidently $\mu = 1$.

The *discharge-head relation* across the constriction can be deduced from the energy equation by assuming that the friction slope is nearly compensated for by the bottom slope. The relation involving the upstream energy head H_o and the critical energy head H_c thus reads

$$H_o = h_o + \frac{Q^2}{2gB^2h_o^2} = \frac{3}{2}\left(\frac{Q^2}{g\mu^2b^2}\right)^{1/3}. \tag{3.16}$$

With $f = Q/(gb^2h_o^3)^{1/2}$ as the constriction Froude number, this can also be expressed as

$$\mu = \left(\frac{3}{2 + \psi^2 f^2}\right)^{3/2}. \tag{3.17}$$

The relation between f, ψ and α is plotted in Figures 3.12(b) to (d) for various values of λ. For any geometry assumed, the parameter f, and thus Q, or h_o may be determined. Also, the effect of a modification of

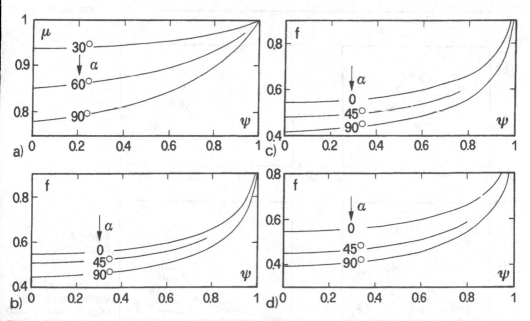

Figure 3.12 Polygonal constriction under *transitional flow* (a) contraction coefficient μ for $\lambda = 1$, relative discharge $f = Q/(gb^2h_o^3)^{1/2}$ for $\lambda = $ b) 0, c) 1, d) ∞

any parameter on the other may be deduced. The effect of the upstream angle on f for constriction rates below $\psi = 0.6$, say, is relatively small.

3.3.3 Subcritical flow

For a relatively large tailwater depth h_t or for a larger value of ψ, the flow is not transitional but *submerged*. Figure 3.13 refers to typical sectional views of both cases. The case with a transition from sub- to supercritical flow, and a hydraulic jump in the tailwater are also plotted.

The contraction coefficient μ for subcritical flow across a constriction depends not only on the approach angle α, the constriction rate ψ and the constriction length λ but in addition on the ratio of flow depths in the contracted and the upstream sections $Z = h_2/h_o$. An expression for μ was determined by Hager (1987). By accounting for the local losses in the expanding tailwater portion a relation may be obtained for the depth ratio $T = h_o/h_t \geq 1$ as a function of the tailwater Froude number $\mathbf{F}_t = Q/(gB^2h_t^3)^{1/2}$. Figure 3.14 refers to results for constriction

Figure 3.13 Surface profiles across a river constriction with subcritical approach flow and ① subcritical tailwater, ② local transition to supercritical flow followed by a hydraulic jump, ③ supercritical tailwater. (. . .) critical depth profile

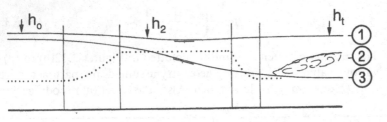

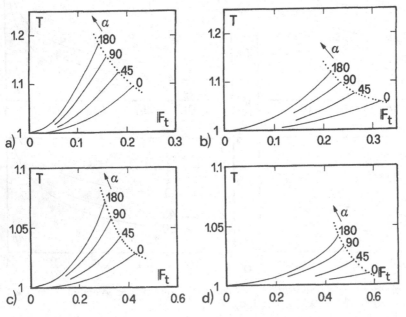

Figure 3.14 Depth ratio $T = h_o/h_t$ as a function of $\mathbf{F}_t = Q/(gB^2h_t^3)^{1/2}$ for various approach angles α and $\psi = $ (a) 1/3, (b) 1/2, (c) 2/3, (d) 5/6

rates between $\psi = 1/3$ and $5/6$. The computations were checked with model observations. The dotted lines refer to the transition between submerged and free flow. According to Figure 3.14, the effects of both α and β on T are relatively small, but the tailwater Froude number $\mathbf{F}_t$ and the constriction rate ψ have a significant effect.

The effect of a *rounded constriction geometry* in plan was considered by Sinniger and Hager (1989). Separation of flow to a contracted width μb is inhibited (i.e. $\mu = 1$) if the radius of curvature upstream from the contracted cross-section (subscript C) is larger than $R_o/B = \mathbf{F}_C^2$ where $\mathbf{F}_C = Q/(gb^2h_C^3)^{1/2}$. For critical flow, i.e. $\mathbf{F}_C = 1$, the minimum radius of curvature is thus $R_o = b$. The effect of flow contraction can be inhibited with a comparably small effort.

3.4 CULVERT

3.4.1 Introduction

A culvert is an artificial hydraulic tunnel structure across a dam or an embankment with either a free surface or a pressurized flow. The design of the culvert is thus closely related to the definition of *flow type* for any relevant discharge and tailwater level. It involves the computation of the head-discharge relation. Culverts are occasionally also used in parallel. At the end of construction, they are either closed with stop logs or a concrete plug, or they are integrated into the bottom outlet (Chapter 6) and the withdrawal structure (Chapter 7). In the following, a generalized approach is presented by which preliminary information may be obtained. Culverts should also be checked with regard to vortex setup in the approach domain, and scour development at the culvert outlet. These aspects will be mentioned also. Figure 3.15 shows a typical culvert as used for a dam.

3.4.2 Hydraulic design

The culvert can have a wide variety of cross-sectional shapes including rectangular, horseshoe or circular. The latter has a circular pipe of diameter D and of length L_d. The bottom slope is S_o, the approach energy head is H_o relative to the inlet invert, and the outlet energy head is H_u relative to the outlet invert. The inlet is often curved with a radius of curvature r_d. Figure 3.16 refers to this simplified culvert structure. Culverts are said to be long if $L_d/D > 10$ to 20, and flow patterns become not as complex as for the short culvert.

Chow (1959) has summarized the main flow types in a culvert (Figure 3.17):

Figure 3.15 Typical culvert structure. Al Massira dam culvert (Morocco) during construction (VAW 51/6/3)

① For $H_o < 1.2D$ and $S_o > S_c$ where S_c is the critical bottom slope (i.e. the slope for which the uniform flow is equal to the critical flow) *critical flow* develops at the inlet.

② For $H_o >$(1.2 to 1.5)D and for free outflow, so-called *gate flow* occurs. The culvert inlet corresponds hydraulically to a gate structure, with unlimited air access from the tailwater into the culvert.

③ For $H_o < 1.2D$ and $S_o < S_c$ subcritical flow develops, with the control either at the culvert outlet, or in the tailwater channel. In

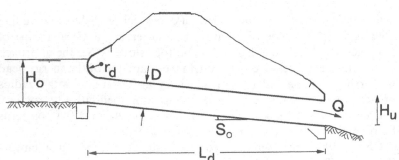

Figure 3.16
Culvert type
structure, notation

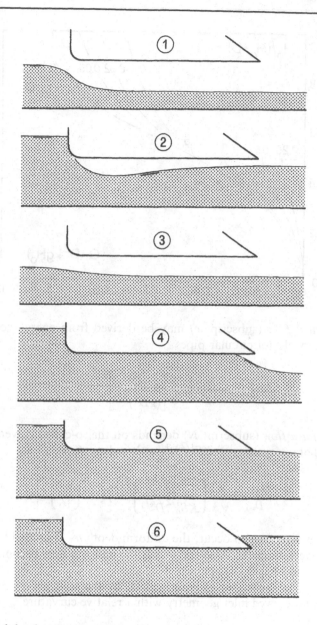

Figure 3.17
Types of flows in a
culvert structure

general, backwater curves dominate this flow type, and *uniform flow*
may be assumed for long culverts.

④ For a tailwater submergence $h_c < H_u < D$ and a small bottom slope
($S_o < S_c$) the culvert is partially submerged.

⑤ For a *pressurized flow* with the outflow depth equal to the culvert
height, the tailwater may become supercritical; and

⑥ For a completely *submerged outlet* ($H_u > D$) and $S_o < S_c$ the flow in
the culvert is pressurized with a subcritical outflow.

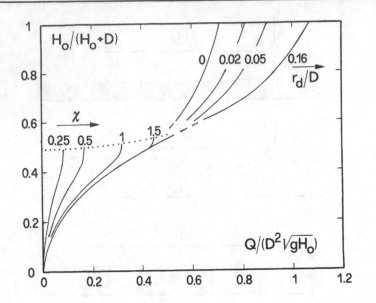

Figure 3.18
Generalized culvert design diagram as a function of roughness characteristics χ from free surface to pressurized flow (Sinniger and Hager 1989)

Critical flow (subscript c) may be derived from energy considerations to yield for circular pipes

$$\frac{H_{oc}}{D} = \frac{5}{3}\left[\frac{Q}{(gD^5)^{1/2}}\right]^{3/5}.$$ (3.18)

Uniform flow (subscript N) depends on the so-called culvert roughness characteristics $\chi = KS_o^{1/2}D^{1/6}g^{-1/2}$ where $K = 1/n$ as

$$\frac{H_{oN}}{D} = \frac{2}{\sqrt{3}}\left(\frac{Q}{KS_o^{1/2}D^{8/3}}\right)^{1/2}\left[1 + \left(\frac{9\chi}{16}\right)^2\right].$$ (3.19)

For uniform flow to occur, the uniform depth h_N is limited to 90% of the culvert diameter. Also for $\chi > 2$, the uniform flow becomes supercritical, and Eq.(3.18) applies.

Gate flow (subscript g) is influenced by the contraction coefficient C_d. For a curved inlet geometry with a relative curvature $\eta_d = r_d/D$

$$C_d = 0.96[1 + 0.50\exp(-15\eta_d)]^{-1}.$$ (3.20)

According to the Bernoulli equation the discharge obtains

$$Q_g = C_d(\pi/4)D^2[2g(H_o - C_dD)]^{1/2}.$$ (3.21)

The increase of discharge as a function of C_d is limited to $\eta_d \leq 0.15$. Rounding the inlet by a radius of $r_d = 0.15D$ yields an inflow without separation from the vertex. The pressure may decrease below the atmospheric pressure, however, and vibrations due to the boundary layer detachment may occur.

Pressurized flow (subscript p) is influenced by the headlosses across the culvert. If $\Sigma\zeta$ represents the sum of all losses, the discharge equation reads

$$Q_p = (\pi/4)D^2[2gH_d/(1+\Sigma\zeta)]^{1/2} \qquad (3.22)$$

where $H_d = H_o + S_oL_d - H_u$ is the head on the culvert. For a reasonably rounded inlet and a straight prismatic pipe, the resistive forces are made up by friction forces, i.e. $\Sigma\zeta = 2\cdot 4^{4/3}gL_d/(K^2D^{4/3})$. Figure 3.18 shows a generalized discharge-head diagram where the critical, the uniform and the gate type flows are accounted for. A simple design is thus possible. It may be applied correspondingly to other cross-sectional shapes.

REFERENCES

ASCE Task Force (1965). Factors influencing flow in large conduits. *Journal Hydraulics Division ASCE* **91**(HY6): 123–152; **92**(HY4): 168–218; **93**(HY3): 181–187.

Barr, D.I.H. (1973). Resistance laws for large conduits. *Water Power* **24**(8): 290–304.

Bray, D.I. and Davar, K.S. (1987). Resistance to flow in gravel bed rivers. *Canadian Journal Civil Engineering* **14**: 77–86; **14**: 857–858.

Chow, V.T. (1959). *Open channel hydraulics*. McGraw-Hill, New York.

Colebrook, C.F. (1958). The flow of water in unlined, lined, and partly lined rock tunnels. *Proc. Institution Civil Engineers* **11**: 103–132; **12**: 523–562.

Colebrook, C.F. and White, C.M. (1937). Experiments with fluid friction in roughened pipes. *Proc. Royal Society* London A **161**: 367–381.

Hager, W.H. (1987). Discharge characteristics of local, discontinuous contractions II. *Journal Hydraulic Research* **25**(2): 197–214.

Hager, W.H. and Dupraz, P.A. (1985). Discharge characteristics of local, discontinuous contractions I. *Journal Hydraulic Research* **23**(5): 421–433.

Hinds, J. (1928). The hydraulic design of flume and syphon transitions. *Trans. ASCE* **92**: 1423–1459.

Jaeger, C. (1949). *Technische Hydraulik*. Birkhäuser, Basel (in German).

Jarrett, R.D. (1984). Hydraulics of high-gradient-streams. *Journal Hydraulic Engineering* **110**(11): 1519–1539; **113**(7): 918–929.

Meyer-Peter, E. and Müller, R. (1948). Formulas for bed-load transport. *2 IAHR Congress* Stockholm **2**: 1–26.

Schröder, R.C.M. (1990). Hydraulische Methoden zur Erfassung von Rauheiten (Hydraulic methods to determine roughnesses). *DVWK Schrift* **92**. Parey, Hamburg and Berlin (in German).

Sinniger, R.O. and Hager, W.H. (1989). *Constructions hydrauliques* (Hydraulic structures). Presses Polytechniques Fédérales, Lausanne (in French).

Smith, C.D. (1967). Simplified design for flume inlets. *Journal Hydraulics Division* ASCE **93**(HY6): 25–34; **94**(HY3): 813–815; **94**(HY4): 1152–1153; **94**(HY6): 1544–1545; **95**(HY4): 1456–1457.

Strickler, A. (1923). Beiträge zur Frage der Geschwindigkeitsformel und der Rauhigkeitszahlen für Ströme, Kanäle und geschlossene Leitungen (Contributions to the question of velocity formula and the roughness numbers for rivers, channels and closed conduits). *Mitteilung* **16**. Amt Wasserwirtschaft, Berne (in German).

Vischer, D. (1987). Das Wasserkraftwerk Clyde am Clutha in Neuseeland. (The hydro power plant Clyde on Clutha river in New Zealand). *Wasserwirtschaft* **77**(2): 63–68 (in German).

Vischer, D.L., Hager, W.H., Casanova, C., Joos, B., Lier, P., Martini, O. (1997). Bypass tunnels to prevent reservoir sedimentation. *19 ICOLD Congress* Florence, **Q74**(R37): 605–624.

Visitors at LG2 cascade spillyway, Hydro-Quebec, Canada (*Forces* 105, 87)

4

Outlet Structures

4.1 FREE FALL

4.1.1 Introduction

Among various types of *dam outlets*, the free fall, the spillway chute, and the spillway cascade are the most prominent structures. These are dealt with in this chapter with regard to the hydraulic approach. Figure 2.9 shows sketches of typical outlet structures.

The *free fall outlet* is not a very common type of outlet structure, mainly due to concerns with impact pressure and scour just beneath the dam. It can be used exclusively at locations with excellent underground conditions. Also, a stilling basin is often provided to create a *water cushion* by which the impact action of the falling jet can be reduced. Thomas (1976) mentions that a unit discharge of $80\,\mathrm{m}^2\mathrm{s}^{-1}$ should not be exceeded and that fall heights over 100 m should receive particular attention. He suggests a fully aerated and steady nappe to counter impact pulsations. Figure 4.1(a) refers to such a dam, with an adversely sloping stilling basin to provide a minimum submergence effect. Figure 4.1(b) shows a view from the tailwater to the Gebidem dam, Switzerland, for a discharge of $30\,\mathrm{m}^3\mathrm{s}^{-1}$.

Free fall outlets can only be applied in relatively wide valleys with excellent geological characteristics. The hydraulic design of the structure involves four main features:

Figure 4.1
(a) Free fall outlet structure at Mratinje dam, former Yugoslavia, (b) Gebidem dam, Valais, Switzerland (*Schweizerischer Wasserwirtschafts-Verband* 1970 **42**:53).

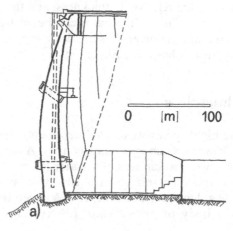

b)

Figure 4.1   *(Continued)*

(a) overflow, (b) nappe flow, (c) impact action, and (d) stilling basin.

In the following, items (b) and (c) are dealt with. The overflow structure is analogous to the standard spillway overflow, except that a small *deflector* discharges the jet in the air instead of a chute. The jet trajectory is then fully determined by the actions of gravity, viscosity, surface tension and air entrainment. The intensity of impinging action mainly depends on the relative jet thickness, and the degrees of turbulence and dispersion. The stilling basin can be regarded as a separate structure whose main characteristics are dealt with in Chapter 5. Some features of the impact basin are presented.

4.1.2   Jet trajectory

Water jets are highly complex to analyse, mainly due to the complex interaction of gravity, viscosity and capillarity. A further complication stems from the two free surfaces of a water jet in air. Currently, such jets may *numerically* be treated in a two-dimensional situation for inviscid flow (e.g. Dias and Tuck, 1991, or Vanden Broeck and Keller, 1986). A large body of *experimental* observations on spillway jets

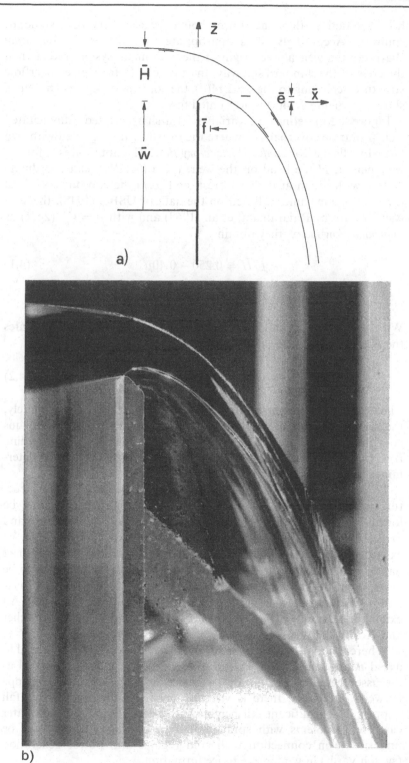

Figure 4.2
(a) Definition of nappe geometry, origin of coordinates at weir crest, (b) outflow nappe.

however defines the exact lower boundary geometry of a standard spillway. Accordingly, it is appropriate to consider the *substitute* sharp-crested weir as the origin of the coordinate system rather than the crest of the standard spillway. In other words, for a given overflow structure according to Figure 2.14(b), the corresponding design weir as shown in Figure 2.14(a) is accounted for.

Figure 4.2(a) defines the variables of a sharp-crested, fully aerated weir (notations overbarred) where the overflow jet is guided with two lateral walls, i.e. for *confined trajectory flow*. The notation is $\bar{w}$ for the weir height, $\bar{H}$ the head on the weir, $(\bar{x}, \bar{z})$ the Cartesian coordinate system with origin at the weir crest, and $(\bar{e}, \bar{f})$ the coordinates of the lower nappe maximum. Based on the data of USBR (1948), the analysis of data by Rajaratnam, et al. (1968) and with $\mu = V_o^2/(2g\bar{H})$ as *approach flow index*, they obtain

$$\bar{f}/\bar{H} = 0.250 - 0.40\mu, \tag{4.1}$$

$$\bar{e}/\bar{H} = 0.112 - 0.40\mu.$$

With $\bar{X} = 1.5(\bar{x}/\bar{H})$ and $\bar{Z} = 3.5(\bar{z}/\bar{H})$ as nondimensional coordinates the *lower nappe profile* simply reads in extension of Eq.(2.2)

$$\bar{Z} = \bar{X} \ln \bar{X}[1 + \frac{1}{6}\bar{X}]. \tag{4.2}$$

The upper nappe geometry has also been investigated. Approximately, the *vertical* jet thickness t_j beyond the lower nappe maximum remains constant $\bar{t}_j/\bar{H} = 0.70$, and the upper jet profile is determined by adding $\bar{t}_j$ to the lower nappe elevation. The impact angle can thus be determined at any elevation below the crest.

To counter *jet vibration*, the offtake portion of the overflow structure is equipped with *nappe splitters* (Figure 4.3). These should be located at the most downstream point of the overflow structure and project through the nappe at design discharge. They should be extended over the lip and burst the nappe. The aeration of the compact nappe also enhances the jet dispersion, and less impact action is to be expected at the stilling basin. According to Schwartz (1964) the maximum *splitter spacing* should be about two-thirds of the fall height. An extensive account on *flow-induced vibrations* is given by Naudascher and Rockwell (1994).

Whereas additional *aeration* of a gated overflow nappe is essential to avoid *nappe oscillation* it is not always needed for a free overfall as discussed here. Because the air has no access below the overflow nappe between gate piers, there is virtually no air restriction for free fall nappes. The significant fall heights and the large quantities of water can give problems with spray, however. The *spray action* can be undesirable in connection with cold regions and power generation (switch yards) in winter due to ice formation.

a)   b)

Figure 4.3   (a) Oscillating nappe, and (b) nappe splitters (Schwartz, 1964)

4.1.3   Jet impact

A jet issued from an overfall structure interacts with the surrounding
air and develops to an *aerated turbulent jet* (Figure 4.4). Depending on
the relative jet thickness, its turbulence degree, and the fall height, the
average air concentration may be small and the impact effect significant, or the jet may have nearly dispersed in the air. The latter case is
certainly not usual for design flow, and free fall nappes have normally
a large impact load. Because the overflow jet is not directed away from
the dam, such as for trajectory jets, a *bottom protection* is needed.
Examples such as the Kariba dam with orifice jets, i.e. that are directed
away from the dam section have experienced significant scour. For free
fall jets, it is thus highly unadvisable to use a plunge pool basin instead
of an impact basin (Chapter 5).

The impact jet characteristics are similar to the wall jet characteristics, involving a turbulent free zone of the approach jet, a
deflection zone in the impact region, and a wall jet zone further downstream. A minimum *water cushion* must be provided to submerge the
impinging jet. Details of such flow are reviewed by Vischer and Hager
(1995).

The time-averaged *maximum bottom pressure* p_M depends on the
ratio of jet thickness t_j to the height t_L of the water cushion (Figure
4.5(a)). It may be approximated with the impact velocity V_1 as

$$p_M/(\rho V_1^2/2) = 7.4(t_j/t_L). \qquad (4.3)$$

The local pressure distribution is of Gaussian type and the *rms-
pressure fluctuations* can be correlated as

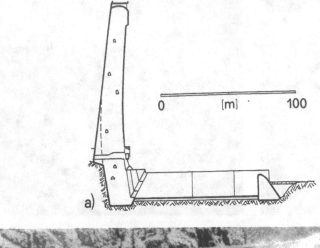

Figure 4.4
Morrow Point dam,
upper Colorado
region (USA),
(a) section and
(b) tailwater view
(Koolgaard and
Chadwick, 1988)

$$\overline{(p'^2)}^{1/2}/(\rho V_1^2/2) = \alpha \qquad (4.4)$$

where the upper and lower extremes are +0.28 and −0.04, respectively.

The stilling basin should have a standard lining as described in Chapter 5. Additional information on basins with an impact jet is currently not available, and detailed *hydraulic modelling* is therefore recommended. It should also be kept in mind that the impact jet divides at the basin bottom and that a significant flow portion can be directed upstream towards the dam (Figure 4.5(b)). Walls must be provided so that scour action is inhibited.

4.2 CHUTE

4.2.1 Hydraulic design

A chute is a sloping open channel made of concrete by which excess discharge is conveyed from the overflow structure to the tailwater. The particular feature of *chute flow* is its high velocity normally above $20 \, \text{ms}^{-1}$, and up to $50 \, \text{ms}^{-1}$. Problems with air entrainment, shock waves, cavitation and abrasion are frequent. To guarantee reliability under extreme flood conditions, the chute must be designed and

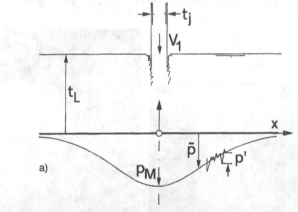

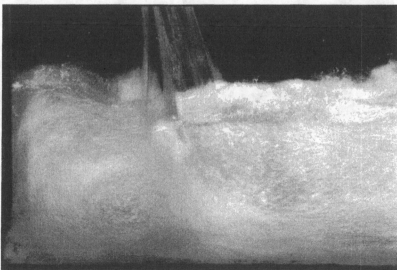

Figure 4.5
Impact action of jet
(a) notation,
(b) typical impact
jet in lab channel

constructed carefully. Endangering a chute can often mean a direct danger to the dam.

The chute is one of the most spectacular portions of a dam, and is a particular challenge for each designer and hydraulic engineer. Chutes are typically used for spillways which are separated from the dam structure. Outstanding examples are located at the Itaipu dam (Brazil/ Paraguay) or the Tarbela dam (Pakistan). Typically, a chute may have a design unit discharge up to $100 \, m^2 s^{-1}$, and the bottom slope can typically vary from 20° to 60°. Usually, walls are vertical or have a slope 1:2, and the chute surface is smooth to inhibit cavitation damage. Particular attention must be paid to concrete joints because even a small offset of less than 10 mm can become the origin of cavitation (Falvey, 1990). The concept of the chute spillway is different from the cascade spillway, where the water is tumbling over a series of steps. In chute flow, minute perturbations of flow are amplified and can be visualized as a *cross-wave*. As an example, the perturbations of the gate piers give rise to so-

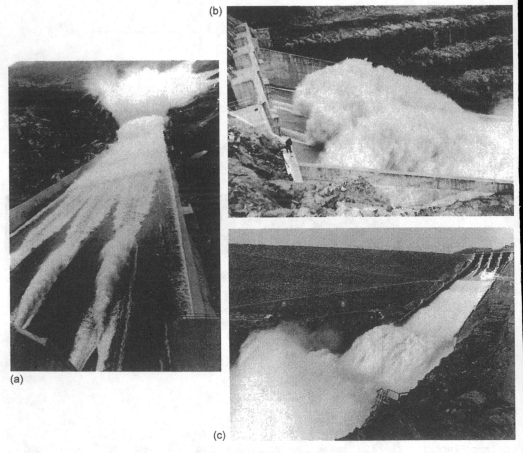

Figure 4.6 Gate pier waves and chute flow at Foz do Areia dam (Brazil) (a) upstream view, (b) side view with roostertails, (c) downstream view with ski jump for $Q = 2500 \, m^3 s^{-1}$ (Courtesy N. Pinto)

called *roostertails*, which can clearly be seen in Figure 4.6 as high standing waves. Due to the low pressure at the rear of pier, air is locally entrained in the chute flow, and traces the way downstream from the origin of perturbation. These waves spread diagonally over the chute and eventually reach the side walls, where wave reflection occurs, associated with a local increase of wall flow depth. Crosswaves should thus be reduced in height for an economic chute design.

The addition of air to the highspeed water flow is effective and economic to counter cavitation damage on chutes. Special aeration devices, so-called *chute aerators* are thus provided at locations where the natural free surface air entrainment does not suffice for the concrete protection. In contrast to air entrainment at chute perturbations (such as roostertails previously discussed), the chute aerator is designed for a certain amount of air discharge, and the air is distributed over the entire chute width. Accordingly, one should clearly distinguish between local perturbations and aerators of chutes for controlling the air entrainment. Then, the freeboard may be kept relatively small and the energy dissipator, located downstream of a chute, experiences no adverse effect.

The hydraulic design of a chute has to account for various items, including:

- the surface air entrainment and the bulking of the air-water mixture,

- the effect of chute aerators, and

- the effect of shockwaves.

These items are discussed in the following sections.

4.2.2   Surface air entrainment

Fast flowing water is known to become white, because air is entrained in the flow (Figure 4.7). Such air entrainment is also referred to as natural, as opposed to so-called forced air entrainment with aerators. For air entrainment to occur, two *conditions* must be satisfied:

1. The flow must be fully turbulent, i.e. the thickness of the boundary layer must be equal to the flow depth, and

2. The kinetic energy of surface eddies must be greater than the surface tension energy.

The latter condition is usually fulfilled in prototype structures. Depending on the degree of turbulence, or the turbulence number, chute flow gets naturally aerated in prototype structures when the velocity is in excess of 10 to $15\,\mathrm{ms}^{-1}$.

Uniform aerated flow may be compared with the conventional uniform flow in channels, except for the equilibrium between tractive and retarding forces for aerated flow. Using the concept of the uniform aerated flow one may distinguish between equilibrium and nonequilibrium reaches along a chute (Figure 4.8).

The growth of a *turbulent boundary layer* downstream from an overflow structure was predicted by Wood, et al. (1983) as

$$\frac{\delta}{x} = 0.0212 \left(\frac{x}{H_s}\right)^{0.11} \left(\frac{x}{k_s}\right)^{-0.10} \tag{4.5}$$

a)   b)

Figure 4.7 (a) Pier waves visible up to first aerator, (b) high concentration air-water flow up to trajectory basin at Italpu dam (Brazil/Paraguay) (Courtesy N. Pinto)

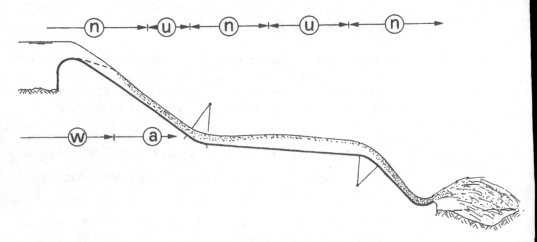

Figure 4.8 Chute flow from overflow structure to energy dissipator, with *n* non-equilibrium and *u* uniform or equilibrium flow reaches. *w* water and *a* air-water flow

where δ is the thickness of the boundary layer, x the longitudinal coordinate measured from the crest, H_s the difference of water elevations between the inception point and the reservoir level, and k_s the equivalent roughness height according to Nikuradse (Figure 4.9). Approximately, this simplifies to

$$\delta/x = 0.02(k_s/H_s)^{0.10}. \tag{4.6}$$

The *inception point* $x = x_i$ is located where the local flow depth $h = h(x)$ is equal to the thickness of boundary layer $\delta = \delta(x)$, i.e. $\delta_i = h_i$. From Eq.(4.6), one may note that increasing the equivalent roughness has the same effect as decreasing the elevation difference. The point of inception can be determined with a drawdown curve. Reinauer and Hager (1996) have presented a simplified analysis that was checked with lab data.

The *uniform aerated flow* involves the air concentration profiles. Based on the classic set of data from Straub and Anderson (1960) collected for chutes sloping between 7.5° and 75°, Wood (1985) was able to plot generalized concentration profiles. If h_{90} is the depth over the chute bottom with 90% air concentration, and y the coordinate perpendicular to the bottom, then the *equilibrium concentration distribution* C_e varies only with the nondimensional depth $Y_{90} = y/h_{90}$, and the chute inclination θ. Figure 4.10(a) shows the corresponding curves based on the Straub and Anderson model observations. These are currently accepted for typical concrete prototype chutes.

Hager (1991) has further analysed the uniform aerated flow and found for the *bottom concentration* $C_o = C_e(Y_{90} = 0)$ when overlooking the concentration boundary layer (Figure 4.10(b),(c))

$$C_o = 1.25\theta^3, \qquad 0 < \theta < 0.70 \ (40°); \tag{4.7}$$
$$C_o = 0.65\sin\theta, \quad 0.70 < \theta < 1.40 \ (80°). \tag{4.8}$$

All concentration curves may be scaled as $c = (C_e - C_o)/(0.90 - C_o)$ and vary only with $y^* = 2(1 - Y_{90})(\sin\theta)^{-1/2}$. Figure 4.11(b) shows the universal concentration profile $c(y^*)$.

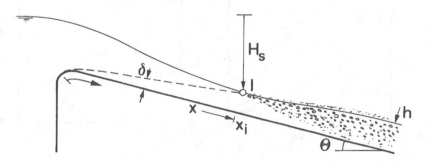

Figure 4.9
Determination of
inception point I

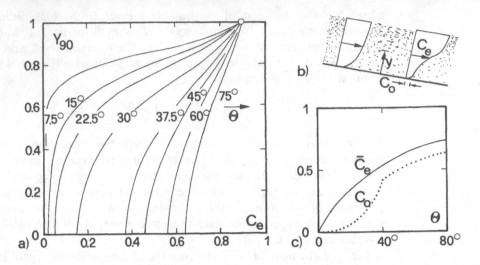

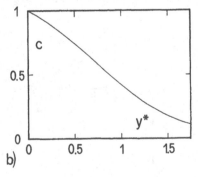

Figure 4.10 (a) Concentration profiles $C_e(Y_{90})$ for *uniform* aerated flow for various chute inclinations θ (Wood, 1985), (b) definition plot and (c) bottom and average air concentrations C_o and $\bar{C}_e$ according to Hager (1991)

The cross-sectional *average equilibrium concentration* $\bar{C}_e$ can be obtained by integrating the concentration curves over the flow thickness as (Figure 4.10(c))

$$\bar{C}_e = 0.75(\sin\theta)^{0.75}. \tag{4.9}$$

Accordingly, the concentration increases almost linearly for usual chute slopes. A generalized approach is presented in Eq.(4.11) below.

The Darcy-Weisbach *friction factor f* for chute flow depends on the relative sand-roughness height and the aspect ratio. If f_w is the factor for water flow and f_e the corresponding factor for equilibrium air-water flow, then the ratio f_e/f_w varies with the average air concentration $\bar{C}_e$, as shown in Figure 4.11(a). For $\bar{C}_e < 20\%$, i.e. for $\theta \leq 10°$ from Eq.(4.9) this effect is negligible, whereas for $\theta > 10°$ the resistance of an air-water mixture is smaller than of the corresponding *pure* water flow.

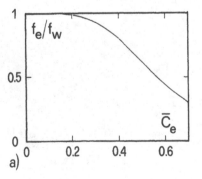

Figure 4.11
Uniform aerated chute flow (a) ratio of friction coefficients for aerated and nonaerated flows, (b) universal concentration profile

The *increase of flow depth* (flow bulkage) due to the air-water mixture varies mainly with the chute roughness characteristics $\eta = [h_w \sin^3\theta/(n^2 g^3)]^{1/4}$ where h_w is the corresponding *pure* water depth and n the Manning friction coefficient. If $Y_{99} = h_{99}/h_w$ is the relative flow depth with 99% air concentration, and b is the chute width then (ICOLD, 1992)

$$Y_{99} = 1 + \frac{1.35\eta}{1 + 2(h_w/b)}. \tag{4.10}$$

The *freeboard* $Y_{99} - 1$ due to the air-water flow depends mainly on η, i.e. significantly on the chute slope and on the wall roughness, and to a small degree on the pure water depth.

Eq.(4.9) is based on the Straub and Anderson data collected in a channel of a definite boundary roughness, i.e. with one roughness value n. For shallow chute flow $(h_w/b \ll 1)$ it may be shown from Eq.(4.10) that

$$\bar{C}_e = 1.35\eta. \tag{4.11}$$

This is a generalization of Eq.(4.9) involving both n and h_w besides the chute slope θ.

For *gradually varied flow,* the effects of backwater curve, and air entrainment may be separated, at least to lowest order of approximation. As shown by ICOLD (1992), one would compute the backwater curves of the *pure* water flow first, and then add the effect of air-water flow, according to Eq.(4.10). An alternative approach was presented by Falvey (1990).

4.2.3 Chute aerator

For uniform aerated flow, and for chute inclinations $\theta < 20°$ to $25°$, the bottom air concentration is below 6 to 8% from Eq.(4.7). Then, the chute bottom is not sufficiently protected against cavitation damage.

Incipient cavitation damage depends on a number of parameters, including the overall quality of the chute surface, the height and number of local surface irregularities, the deviation of the prismatic straight chute geometry, the degree of turbulence and the local flow velocity. For average velocities up to $30 \, ms^{-1}$, a bottom air concentration of 6 to 8% is known to be sufficient for cavitation protection. However, if the quality of chute is lower than previously described, or the hydraulic conditions are more severe, higher air concentrations of up to 10 or 15% are required. The value of 20° for the chute inclination is thus an order for *uniform* aerated chute flow. For accelerating flows, or for flows departing from the standard chute flow, a different approach has to be followed.

Among various methods to counter *cavitation damage*, the chute aerator is the most economical method in dam hydraulics. The flow close to the aerator may be divided into four zones as shown in Figure 4.12 (Volkart and Rutschmann, 1984):

① approach flow usually with a surface aeration,

② transition zone deflected by a ramp,

③ bottom aeration behind the offset by an aeration device, and

④ bottom aeration with gradually varied aerated flow.

A highly turbulent spray develops in the cavity downstream of the aerator, in which subatmospheric pressure exists. The air supply is provided from the atmosphere by an air intake and supply channels. The air is mixed in the chute flow along the lower nappe, from the offtake at the ramp to the impinging point, where a significant bottom pressure increase occurs (Figure 4.12). An aerator is known to increase the local turbulence that enhances the nappe air entrainment (Ervine, et al., 1995).

According to Vischer, et al. (1982), a chute aerator should include either a *ramp*, or an *offset*, or even a combination of the two. Grooves are to be inhibited due to problems with aerator submergence. An aerator should provide an approximately *uniform* distribution of air over the chute width. Figure 4.13 shows two type chute aerators without, and with an air manifold. The latter is suited for wide chutes, and yields a nearly perfect air distribution with an almost plane chute flow.

The *design* of an aerator involves a large number of parameters, such as the driving pressure Δp between the atmosphere and the cavity pressure, the air discharge Q_a, the approach velocity V, the approach flow depth h, the chute slope θ, the deflector slope α, the heights of ramp t_r and offset (step) t_s, and the jet length L_j measured up to the point of maximum bottom pressure (Figure 4.14).

The *air entrainment rate* $\beta = Q_a/Q$ with Q as water discharge depends mainly on the approach Froude number $\mathbf{F} = V/(gh)^{1/2}$ and the *Euler number* $\mathbf{E} = V^2(\Delta p/\rho)^{-1}$ with ρ as air density. Usually, effects of viscosity, surface tension and turbulence can be dropped for prototype concrete chutes, and one may write

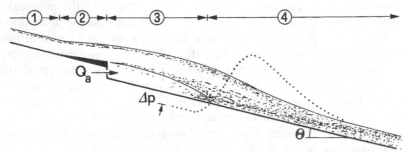

Figure 4.12
Flow zones in vicinity of chute aerator, (...) bottom pressure curve

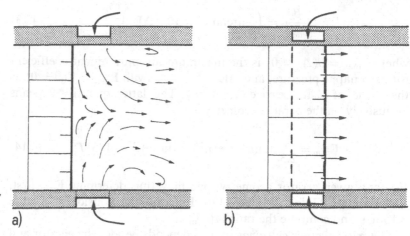

Figure 4.13
Air supply systems
for chute aeration
(a) without, and
(b) with distribution
duct (Vischer, et al.,
1982).

$$\beta = f(\mathbf{F}, \mathbf{E}, \theta, \alpha, T_r, T_s) \qquad (4.12)$$

where $T_r = t_r/h$, $T_s = t_s/h$ are the relative ramp and step heights. The
effect of the Euler number may be split when referring to the maximum
air rate β_{max} as (Rutschmann and Hager, 1990a)

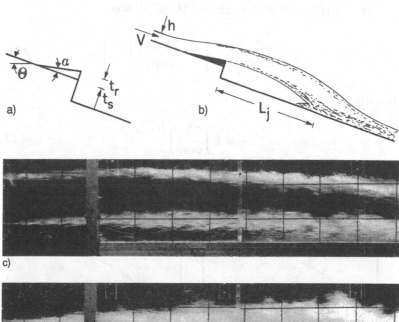

Figure 4.14
Side view of chute
aerator
(a) geometry, and
(b) flow features.
Photographs of lab
flows for
(c) $\mathbf{F} = 5.6, \beta = \beta_{max}$,
(d) $\mathbf{F} = 10.9$, $\beta = 0$

$$\frac{\beta}{\beta_{max}} = \left(\frac{2}{\pi} \arctan(3 \times 10^{-3} \Delta E)\right)^{0.7} \qquad (4.13)$$

where $\beta_{max} = \beta(\Delta p = 0)$ is the maximum air entrainment coefficient for zero nappe pressure, and $\Delta E = E - E_{min}$ with $E_{min} = E(\beta = 0)$ as the value of E for zero entrainment. The latter quantity depends exclusively on the aerator geometry as

$$10^{-3} \times E_{min} = \frac{1}{2.3}(\tan\alpha)^{1.15}\exp[1.15(\tan\theta)^2] + (1/3)T_s^2. \qquad (4.14)$$

For any given aerator geometry, one may thus determine E_{min} first, then evaluate ΔE for all values of the Euler number considered (Figure 4.15(a)), and compute the ratios β/β_{max}.

The *maximum air entrainment* β_{max} depends besides the aerator and chute geometries also on the approach Froude number F. For a long approach chute where nearly uniform aerated flow exists, β_{max} increases with the relative jet length $\lambda_j = L_j/h$ as

$$\beta_{max} = 0.030(\lambda_j - 5) \quad \text{for } F > 6. \qquad (4.15)$$

The length of jet was determined for *ramp aerators* with $T = (t_r + t_s)/h$ as (Rutschmann and Hager, 1990a)

$$\lambda_j = \frac{\bar{\alpha}F^2}{\cos\theta}\left\{1 + \left[1 + \frac{2T\cos\theta}{(\bar{\alpha}F)^2}\right]^{1/2}\right\}, \qquad (4.16)$$

and for *offset aerators* ($\bar{\alpha} = 0$)

$$\lambda_j = F\left(\frac{2T_s}{\cos\theta}\right)^{1/2}. \qquad (4.17)$$

The offset angle correction due to non-hydrostatic pressure distribution is (Figure 4.15(b))

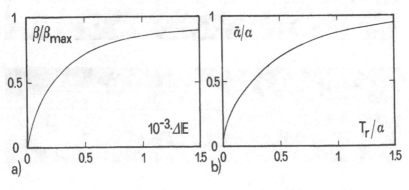

a)   b)

Figure 4.15
Ratio (a) β/β_{max} from Eq.(4.13) and (b) $\bar{\alpha}/\alpha$ from Eq.(4.18)

$$\bar{\alpha}/\alpha = [\text{Tanh}\ (T_r/\alpha)]^{1/2}. \tag{4.18}$$

The determination of the air supply discharge Q_a is thus straightforward, provided $T < 1$, $\tan \alpha < 0.15$, $\tan \theta < 1.5$, and $6 < \mathbf{F} < 12$. The previous equations were derived from limited experimental data, and should be checked with suitable hydraulic models for important chutes. The internal structure of aerator flow was analysed by Chanson (1989a), and Ervine, et al. (1995).

The design of the *air supply system* depends on the ratio Φ of supply duct area A_a and chute flow area (bh) as (Rutschmann and Hager, 1990b)

$$\Phi = \frac{A_a}{bh} \left(0.43(1 + 2\xi_{\text{tot}}) \frac{\rho_a}{\rho_w} \right)^{-1/2} \tag{4.19}$$

where ξ_{tot} is the total head loss coefficient of the air supply system referred to the lateral entrance air velocity V_a and ρ_a/ρ_w the air-water density ratio. The *air entrainment ratio* β may also be expressed as

$$\beta = \frac{0.03(\lambda_j - 5)}{1 + 0.0085 \frac{\mathbf{F}\lambda_{\text{max}}}{\Phi}}. \tag{4.20}$$

Based on an analysis of *optimum aerator efficiency*, the section ratio is

$$\Phi_{\text{opt}} = \frac{1}{2}(0.0085\mathbf{F}\lambda_{\text{max}}), \tag{4.21}$$

corresponding to

$$\beta_{\text{opt}} = (1/3)\beta_{\text{max}}. \tag{4.22}$$

The optimum air entrainment ratio is thus one third of the maximum entrainment ratio as given in Eq.(4.15). The air supply system should be designed for maximum air velocities of some $50\,\text{ms}^{-1}$.

The *optimum aerator height* $T_{\text{opt}} = [(t_r + t_s)/h]_{\text{opt}}$ in terms of chute freeboard required for $0.4 < \mathbf{F}/\Phi < 120$ should be

$$T_{\text{opt}} = \frac{22\cos^2\theta}{\alpha\mathbf{F}^4}(\Phi/\mathbf{F})^{0.11}. \tag{4.23}$$

If the simultaneous optimum (superscript *) in both parameters Φ and T is sought, the analysis yields (Rutschmann and Hager, 1990b)

$$T^*_{\text{opt}} = \frac{15\cos^2\theta}{\alpha\mathbf{F}^4}. \tag{4.24}$$

The optimum ramp height thus varies significantly with the approach Froude number $\mathbf{F}$, inversely with the ramp inclination α, and with the chute inclination θ.

A reconsideration of all model and prototype data available by 1989 based on the previous approach has proved the reliability of the present model for aerator design. A more recent state of the art was presented by Kells and Smith (1991). A full account on air entrainment is also given by Wood (1991).

Correctly designed aerators should never submerge, i.e. the air supply pipe should not be cut by stagnant water and inhibit the entrainment of air. Eq.(4.15) requires that $\lambda_j > 5$ in order that β_{max} is positive. Aerators prone to submergence are either on a bottom slope which is too small, or have a ramp, which is too steep such that a portion of flow recirculates upon impingment of the jet on the down-stream chute. Currently, little information on the *aerator submergence* is available.

The *air detrainment* was analysed in detail by Chanson (1989b). A simplified approach follows here. Due to the buoyancy of air bubbles, the concentration downstream of an aerator is reduced from the peak value to the equilibrium value. With u_r as bubble rise velocity, one has for the local change of air flow q_a per unit width

$$\frac{\mathrm{d}q_a}{\mathrm{d}x} = -(\bar{C} - \bar{C}_e)u_r\cos\theta. \qquad (4.25)$$

With the continuity equation $(1 - \bar{C})q_a = \bar{C}q_w$ the local change of average concentration is

Figure 4.16
Chute flow with aerators at Foz do Areia dam (Brazil), Courtesy N. Pinto

$$\frac{d\bar{C}}{dx} = \frac{u_r \cos\theta}{q_w}(\bar{C}_e - \bar{C})(1 - \bar{C})^2. \tag{4.26}$$

This equation must be integrated subject to the boundary condition $\bar{C}(0) = \bar{C}_0$, where $\bar{C}_0$ is composed of the aerator approach concentration plus the air concentration $\bar{C}_*$ provided by the aerator. If $\bar{C}_0$ is negligible, then $\bar{C}(0) = \bar{C}_*$ with $\bar{C}_* = \beta/(1+\beta)$ where β is the air entrainment rate. With the substitutions

$$\bar{c} = \frac{1 - \dfrac{1 - \bar{C}_e}{1 - \bar{C}}}{1 - \dfrac{1 - \bar{C}_e}{1 - \bar{C}_*}}, X_* = \frac{1 - \bar{C}_e}{(1 - \bar{C}_e)^{-1} - (1 - \bar{C}_*)^{-1}} \cdot \frac{u_r \cos\theta}{q_w} x \qquad \begin{matrix} (4.27) \\ (4.28) \end{matrix}$$

for relative concentration and dimensionless location, the solution of Eq.(4.26) reads

$$X_* = -[1 - \bar{c} + \ln \bar{c}]. \tag{4.29}$$

The bubble rise velocity is nearly $u_r = 0.16\,\mathrm{ms}^{-1}$ for scale models and $0.40\,\mathrm{ms}^{-1}$ for prototype chutes, due to the different bubble diameters.

Downstream from the aerator, the air concentration is redistributed and tends asymptotically to the concentration of the uniform aerated flow. The local concentration can be predicted with Eq.(4.29), and a value of $\bar{C} = 30\%$ should not be passed because it secures a bottom air concentration of some 6 to 8%. The *design* of an aerator system is quite laborious, and the indications of Falvey (1990) can be helpful. Locations where aerators are optimally arranged are at breaks of the bottom slope. There are no problems when the slope is increasing but aerator submergence must be carefully checked at locations where the slope decreases. Figure 4.17 shows a typical example of ICOLD (1992).

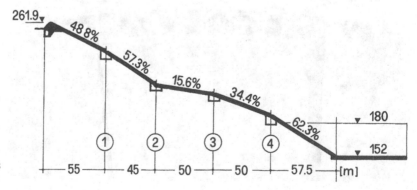

Figure 4.17
Longitudinal chute profile with aerators (ICOLD, 1992)

4.2.4 Shockwaves

Shockwaves or *crosswaves* are a surface pattern in a supercritical flow
along which perturbations are propagated. These waves are often
steady, when the flow remains temporally constant. They can easily
be noticed and they direct to the source of flow perturbation. Crosswaves are running obliquely towards the chute walls and are causing
local maxima of the wall flow profile. Also, a flow becomes nonuni-
form with *local flow extrema* due to the shocks. Figure 4.18 shows an
example with shocks.

The *origin of shocks* is any perturbation of a supercritical flow, i.e.
any change from the uniform chute flow. Typical examples of perturbations in dam hydraulics are:

- chute curves,

- changes of chute slope,

- chute contractions and expansions, or

- chute junctions.

A trajectory bucket induces shocks due to the change of slope over the
bucket. The shockwaves are often visible for small discharges but they
are blurred for larger flows due to the turbulence effect. A shock is not
a real concern for the chute, except that the freeboard must be
increased and that a stilling basin may experience asymmetric
approach flow. A discussion on the basic phenomena, and means, to
reduce shockwaves therefore seems appropriate.

The *basic phenomenon* of a shockwave occurs in the horizontal,
smooth, wide and rectangular channel with an approach flow depth

Figure 4.18
Shockwave at the
rear of gate piers,
Gardiner dam,
Canada (Toth,
1968). Note truck for
comparison of scale

h_1, and an approach velocity V_1. The approach *Froude number* is thus equal to the ratio of approach velocity to approach celerity $F_1 = V_1/(gh_1)^{1/2}$. The approach flow (subscript 1) is referred to as supercritical for $F_1 > 1$, and subcritical otherwise. Usually a subdivision into transcritical ($0.7 < F_1 < 1.5$), and supercritical flows is made, and flows with $F_1 > 3$ are referred to as hypercritical. This discussion includes supercritical flows. Transcritical flows are excluded because of the effect of streamline curvature, and the formation of weak jumps.

Figure 4.19 shows the *abrupt wall deflection*. At the deflection point P, the wall is turned by the angle θ. Because perturbations in supercritical flow are carried downstream only, a perturbation front with a shock angle β is formed. The shock originates at point P, and all streamlines follow the walls upstream and downstream of the shockfront.

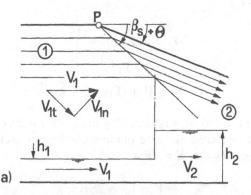

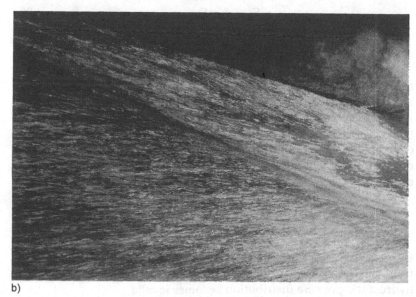

Figure 4.19
Abrupt wall
deflection
(a) notation,
(b) photograph of
shockfront.

Of engineering relevance are the ratio of flow depths h_1 and h_2 in front and behind the shock, the shock angle β and the Froude number F_2 beyond the shock. Based on the momentum equations applied parallel and perpendicular to the shock front yields (e.g. Chow, 1959)

$$Y = \frac{1}{2}[(1 + 8F_1^2 \sin^2 \beta)^{1/2} - 1], \qquad (4.30)$$

$$Y = \frac{\tan \beta}{\tan(\beta - \theta)}, \qquad (4.31)$$

$$F_2^2 = Y^{-1}[F_1^2 - (2Y)^{-1}(Y-1)(Y+1)^2]. \qquad (4.32)$$

Here $Y = h_2/h_1$ is the depth ratio with subscript "1" referring to the unperturbed and "2" to the perturbed flows (Figure 4.19a). For given values of h_1, V_1 and θ one may solve for the unknowns h_2, V_2 and β. An explicit approach for large Froude numbers with $F_1 \sin \beta > 1$ yields (ICOLD, 1992)

$$Y = \sqrt{2} F_1 \sin \beta - \frac{1}{2}, \qquad (4.33)$$

$$\beta = \theta + F_1^{-1}, \qquad (4.34)$$

$$F_2/F_1 = [1 + (F_1 \sin \beta / \sqrt{2})]^{-1}. \qquad (4.35)$$

When the sine function is replaced by the arcus function for the small values of θ considered all three parameters depend exclusively on the so-called *shock number* $S = \theta F$ as

$$Y = 1 + \sqrt{2} S_1, \qquad (4.36)$$

$$\beta F_1 = 1 + S_1, \qquad (4.37)$$

$$F_2/F_1 = (1 + S_1/\sqrt{2})^{-1}. \qquad (4.38)$$

According to Figure 4.19(a) the *wall flow profile* $h_w(x)$ increases abruptly at point P from h_1 to h_2. Schwalt and Hager (1992) have analysed the flow around an abrupt wall deflection and established a generalized surface geometry. Figure 4.20 shows the generalized wall (subscript w) profile as $\gamma_w(X)$ where $X = x/(h_1 F_1)$ is the longitudinal coordinate measured from P and $\gamma_w = (h_w - h_1)/(h_M - h_1)$ is the dimensionless wall height. The *maximum wall flow depth* h_M depends exclusively on the approach shock number S_1 as

$$Y_M = \frac{h_M}{h_1} = 1 + \sqrt{2} S_1 [1 + (1/4) S_1]. \qquad (4.39)$$

Compared to Eq.(4.36), the latter equation has a second order correction which is significant for $S_1 > 0.5$. Then the linear theory based on hydrostatic pressure distribution becomes invalid.

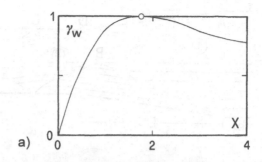

a)

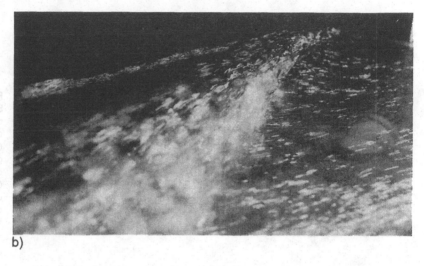

b)

Figure 4.20
(a) Normalised wall profile $\gamma_w(X)$ due to abrupt wall deflection, according to Schwalt and Hager (1992),
(b) shockwave due to abrupt wall deflection for $S_1 = 0.8$, view against flow

The *channel contraction* can be regarded as an application of the abrupt wall deflection. Figure 4.21(a) defines the geometry of the straight-walled structure, with point A as shock origin, point B as shock crossing, point C as wall impingement, and point D as contraction end. At the latter point, a so-called negative wall deflection is seen to occur, with a smooth formation of expansion waves. The approach applies to a tailwater width b_3 larger than 25% of b_1.

Ippen and Dawson (1951) have considered supercritical contraction flow. They introduced a design procedure involving *wave interference* (Figure 4.21(b)). The positive wave originating from point A is supposed to interfere with the wave originating from point D, of which the height is equal but the sign negative. Accordingly, for wave interference to occur, point C must be shifted to point D. The hydraulic condition was determined in ICOLD (1992) and reads for $\theta < 10°$

$$\arctan\theta = \left(\frac{b_1}{b_3} - 1\right) \frac{1}{2\mathbf{F}_1}. \tag{4.40}$$

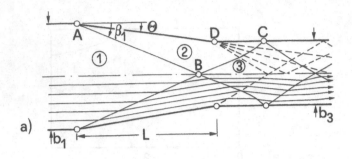

a)

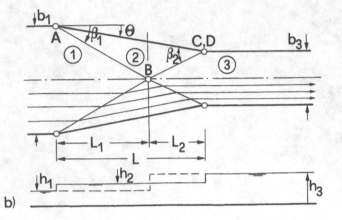

b)

Figure 4.21
Straight-wall channel contraction (a) definition of geometry and off-design flow, (b) design according to Ippen and Dawson (1951) with plan (top) and section (bottom)

The disadvantage of wave interference is that it may be satisfied for one discharge only. Also, Ippen and Dawson's design principle was proved to be in error (Figure 4.22).

Chute Expansion

A channel expansion may typically occur at locations where chutes merge, such as on spillways behind gate piers, or at stilling basins with various parallel chutes subdivided with separation walls. Such flow may also occur at junction structures of small junction angle with only one branch in operation, as well as in bottom outlets. Expanding chutes may be geometrically designed either with an abrupt expansion, or with a Rouse-type expansion.

Rouse et al. (1951) have proposed a continuous expansion function, with a reversed transition to inhibit shockwaves in the tailwater chute. The proposition was checked by Mazumder and Hager (1993) and they found acceptable results also for a 50% transition length as compared with the original Rouse design.

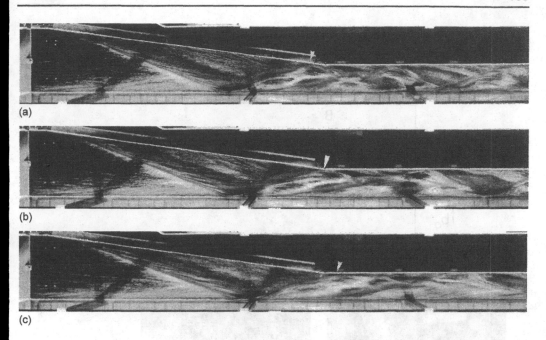

Figure 4.22 Interference principle in contraction with shockfront directed (a) slightly upstream, (b) exactly at and (c) slightly downstream of contraction end point (Reinauer and Hager, 1998)

The *abrupt channel expansion* performs so well that it can often be applied. Figure 4.23 shows the expansion geometry in the horizontal rectangular channel, and defines the axial (subscript a) and wall (subscript w) flow profiles. The scaling lengths are b_o and h_o, and $\beta_e = b_u/b_o, Y_a = h_a/h_o,\ Y_w = h_w/h_o, X = x/(b_o F_o)$ are the parameters for the width ratio, the axial and the wall profile as functions of the relative streamwise coordinate. The free surface profiles determined from extended model studies read (Hager and Mazumder, 1992)

$$Y_a = 0.2 + 0.8\exp(-X^2), \qquad (4.41)$$

$$Y_w = Y_{wM} \cdot \tau\exp(1 - \tau), \qquad (4.42)$$

where the *maximum wall flow depth* (subscript M) for $1.8 < \beta_e < 6$ is

$$Y_{wM} = 1.27\beta_e^{-0.4}. \qquad (4.43)$$

According to Eq.(4.43) the maximum wall depth height h_{wM} depends exclusively on β_e and is always smaller than h_o. The coordinate τ depends on X and β_e as

$$\tau = \frac{X - (1/6)(\beta_e - 1)}{0.52\beta_e^{0.86} - (1/6)(\beta_e - 1)}. \qquad (4.44)$$

The formation of a hydraulic jump downstream from the abrupt expansion is discussed in Chapter 5.

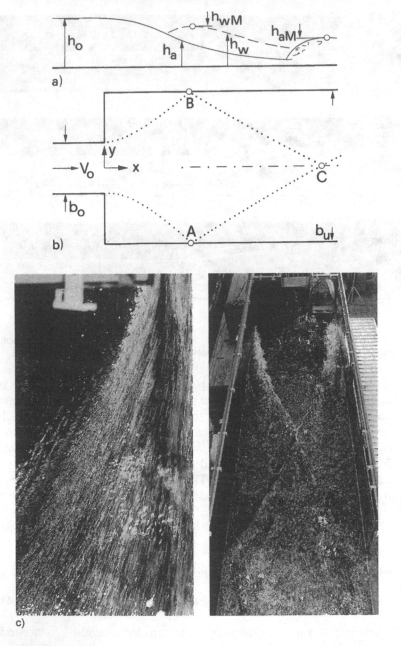

Figure 4.23
Definition of abrupt
channel expansion
(a) section with (—)
axial profile, (---)
wall profile, (b) plan
view with (. . .)
shocks, (c) photos
of abrupt expansion
flow

Chute Bend

Supercritical bend flow is quite common in chutes. A definition plot is
Figure 4.24(a). According to Knapp (1951) the *shock angle* β_s depends
on the relative curvature $\rho_a = b/R_a$ with b as channel width and R_a as
average radius of curvature, and the elementary shock angle
$\sin\theta \cong \tan\theta = \mathbf{F}_o^{-1}$ as

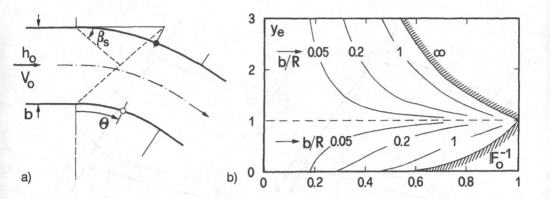

Figure 4.24 Supercritical flow in chute bend (a) definition of flow, (b) extreme flow depths

$$\tan\beta_s = \frac{(b/R_a)\mathbf{F}_o}{(1 + 2b/R_a)} \cong (b/R_a)\mathbf{F}_o. \qquad (4.45)$$

Because the energy loss across a weak shock is small, the energy equation may be applied for the extreme (subscript e) wave heights $y_e = h_e/h_o$ along the outer and inner bend walls

$$y_e = [1 \pm \frac{1}{2}(b/R_a)\mathbf{F}_o^2]^2. \qquad (4.46)$$

This relation is plotted in Figure 4.24(b).

Reinauer and Hager (1997) have analysed *strong bend flow* by selected experiments. The approach of Knapp (1951) was confirmed for small relative curvature, and extended for large curvature. The study dealt with circular bends in rectangular horizontal and smooth chutes. The governing parameter of bend flow is the *bend number* $\mathbf{B}_o = (b/R_a)^{1/2}\mathbf{F}_o$. Figure 4.25 shows typical surfaces across a bend for $\mathbf{B}_o = 1.5$ and 2.3. For Figure 4.25(a) the bend flow is transitional between weak and strong bend flow, and case $\mathbf{B}_o = 2.3$ corresponds to strong bend flow (Figure 4.25(b)). For weak bend flow, the surface is continuous and the transverse sections are nearly trapezoidal. *Separation* from the inner bend wall is locally developed. For strong bend flow, the flow is fully separated from the inner bend wall, and the transverse surface profile is nearly triangular. Also, the shockwave may break.

Of particular *design interest* are the maximum wall depth h_M, and the corresponding location θ_M. Based on detailed observations the results for weak and strong bend flow are, respectively,

$$Z_M = 0.40\mathbf{B}_o^2 \quad , \mathbf{B}_o \leq 1.5; \qquad (4.47)$$
$$Z_M = 0.60\mathbf{B}_o \quad , \mathbf{B}_o > 1.5. \qquad (4.48)$$

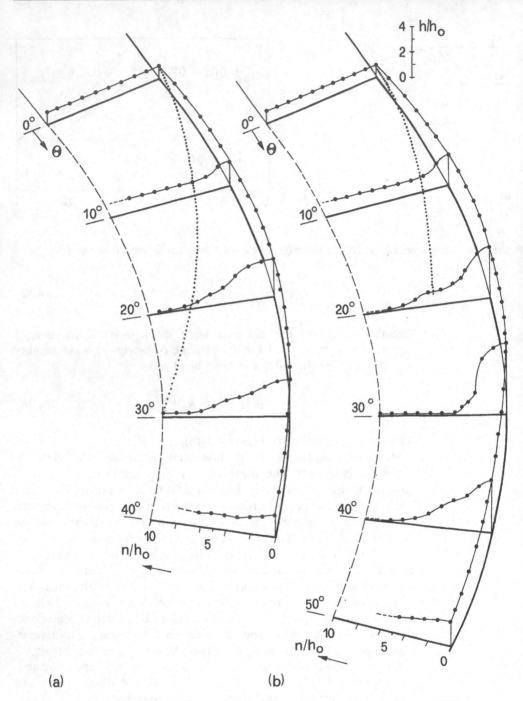

Figure 4.25 Surface of bend flow (a) $B_o = 1.5$, (b) $B_o = 2.3$ (Reinauer and Hager, 1997)

Herein $Z_M = (h_M/h_o)^{1/2} - 1$, and one may note that in a straight chute
($B_o = 0$), the free surface remains undisturbed ($Z_M = 0$). For $B_o \leq 1.5$

the result is basically in agreement with that of Knapp, whereas a linear dependence exists for strong bend flow.

The bend angle θ_M is more difficult to determine, and the data have a larger scatter, therefore. Figure 4.26(b) shows that $\tan\theta_M$ varies with $(b/R_a)\mathbf{F}_o = (b/R_a)^{1/2}\mathbf{B}_o$, which is slightly different from the bend number. The experiments may be approximated as

$$\tan\theta_M = (b/R_a)\mathbf{F}_o, \qquad \text{for } (b/R_a)\mathbf{F}_o \le 0.35; \qquad (4.49)$$

$$\tan\theta_M = 0.60[(b/R_a)\mathbf{F}_o]^{1/2}, \qquad \text{for } (b/R_a)\mathbf{F}_o > 0.35. \qquad (4.50)$$

The minimum wave characteristics, and the wall profiles have also been analysed but are not presented here. More interestingly, the *velocity distribution* is almost tangential with the absolute value close to the approach velocity V_o throughout the main body of bend flow and practically without a radial component. Figure 4.27 shows the development of a curved shock wave, and Figure 4.28 refers to views at the separated bend flow.

Currently, two aspects of supercritical bend flow have not yet received attention: (1) the effect of bottom slope, and (2) the choking conditions. The latter item is extremely important and has to be carefully analysed in sufficiently large hydraulic models. If a chute bend would have been designed for supercritical flow and choking occurs, the flow breaks down and may lead to dangerous and highly unstable *hydraulic jumps*. Typically, choking in horizontal chute bends occurs for Froude numbers below 2 to 4, depending on the curvature and the bottom slope. The effect of bottom slope on the wave maximum is small in analogy to chute contractions.

Shockwaves in tunnel spillways may lead to dangerous *surging phenomena* with abrupt transitions from free surface to pressurized pipe flow. Tunnels with bends have to be designed sufficiently large

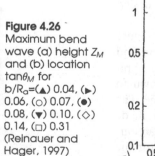

Figure 4.26 Maximum bend wave (a) height Z_M and (b) location $\tan\theta_M$ for b/R_a=(▲) 0.04, (▶) 0.06, (○) 0.07, (●) 0.08, (▼) 0.10, (◇) 0.14, (□) 0.31 (Reinauer and Hager, 1997)

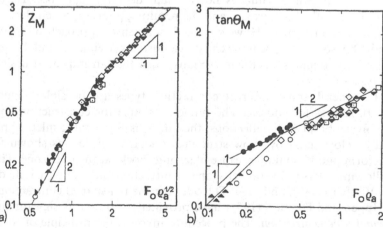

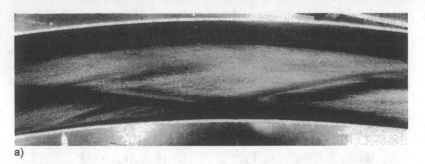

Figure 4.27
Development of
curved shockwave
for $b/R_o = 0.07$ and
$F_o = $ (a) 2, (b) 8

to guarantee free surface flow for all discharges and with *co-flowing air* above the water flow. It is thus important that the two-phase air-water flow is stratified. Figure 4.29 shows tunnel flow and indicates the large transverse surface slopes that may provoke choking and surging. The slope of a tunnel with a supercritical flow, and the tailwater elevation at the tunnel outlet should be chosen such that hydraulic jumps cannot be developed for all flows.

Chute contraction

A contraction in a chute is not a usual design mainly because of problems with shock waves, and the resulting poor approach flow to the energy dissipator. However, the fundamental approach of Ippen and Dawson (1951) was recently extended by Reinauer and Hager (1998) for funnel-shaped contractions, such that an improved design is available.

In general, three different contraction types are possible (Figure 4.30): funnel, fan, nozzle. The funnel is simple for construction and involves abrupt wall deflections, the fan is used for chute inlets, typically below small overflow structures and the nozzle was shown to perform poorly, mainly because of strong shocks at the inlet zone. The following refers to the funnel-shaped contraction and reference is made to Vischer (1988), and ICOLD (1992) for the fan-shaped contraction.

Ippen and Dawson (1951) have provided a rational design for the *funnel type contraction*. The procedure involves the principle of wave

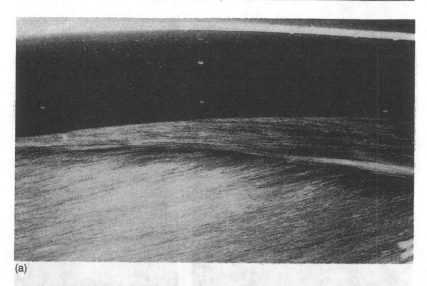

(a)

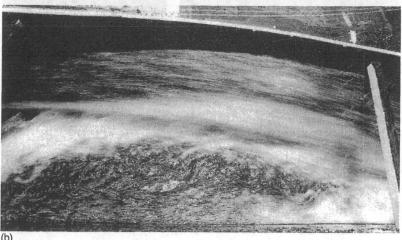

(b)

Figure 4.28
Views at separated
bend flows for
$b/R_a = 0.14$ and
$F_o = 4.5$. Plan views
(a) upper and
(b) lower portions,
views (c) in and
(d) against flow
direction

interference at the contraction end. Figure 4.21 refers to such flow, in
which the contraction angle θ, and the contraction rate b_3/b_1 is chosen
such as the positive shock originating at the contraction point A is
directed to the contraction end point D, from which a negative shock
of equal intensity is interfering, and an undisturbed tailwater flow
should result. Based on detailed observations, this interference principle adopted from optics was shown to fail in chute flow, mainly
because the shockwaves are of finite width and streamline curvature
effects are significant (Figure 4.22).

Figure 4.31 shows the flow pattern in a *plane funnel-shaped contraction* with the approach width b_o and the end width b_e. At the contraction point A a shock of angle β_1 is generated due to the abrupt wall
deflection. The shock is propagated to the channel axis (point B) and
continues to the wall to impinge at point C. At the contraction end

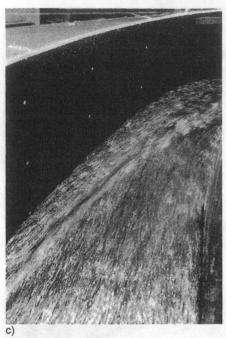
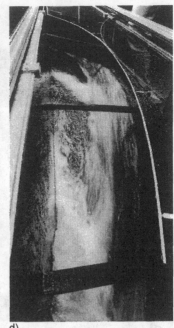

c)   d)

Figure 4.28
(Continued)

point E, a negative wave forms and a complicated wave pattern is set up in the tailwater channel. The profiles in the axis (subscript a) and along the wall (subscript w) are relevant.

Three shock waves can be distinguished:

- wave 1 of height h_1 downstream of point A,

- wave 2 of height h_2 in the axis close to point B, and

- wave 3 of height h_3 along the wall usually downstream of point E.

The locations of the wave maxima are x_1, x_2, and x_3. Downstream of wave 1 the wall flow depth h_p is nearly constant and the flow depth at point E is h_e. The velocity across the shocks through the entire contraction is nearly constant and equal to the approach velocity V_o.

The significant parameters of supercritical contraction flow are the heights and locations of waves 1 to 3, because all waves further in the tailwater are lower. For waves 1 and 2, the width ratio $\beta_e = b_e/b_o$ is insignificant and the governing parameters are approach shock number $S_o = \theta F_o$ in analogy to the abrupt wall deflection, and bottom inclination $\alpha[°]$.

Based on extensive observations, the *wave heights* obtain

$$Wave\ 1 \quad Y_1 = h_1/h_o = (1 + 2^{-1/2}S_o)^2, \qquad (4.51)$$

$$Wave\ 2 \quad Y_2 = h_2/h_o = (1 + 2^{1/2}S_o)^2, \qquad (4.52)$$

$$Wave\ 3 \quad Y_3 = h_3/h_o = \beta_e^{-1} - 0.2\alpha^{0.6} + 1.8S_o^{1/2}. \qquad (4.53)$$

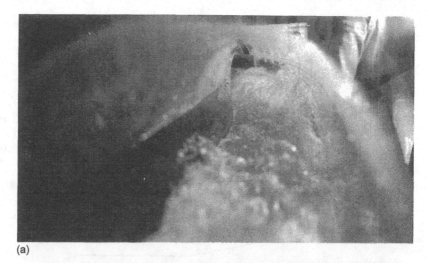

(a)

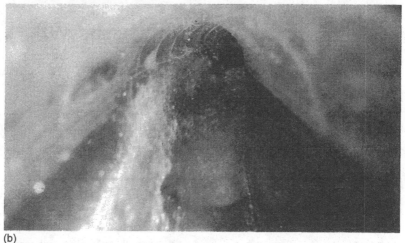

Figure 4.29
Supercritical flow
in tunnel spillway
(a) upstream and
(b) downstream
view with a shock
along the outer
tunnel wall

(b)

Note that the first order approximation of Eq.(4.51) is equal to Eq.(4.36) based on the shallow water assumptions. Also, both waves 1 and 2 are independent of chute slope, at least up to 45°. For wave 3, an increasing slope reduces the wave height, and this result may be transmitted to supercritical flow in general. The wave positions were also determined.

To reduce a shockwave generated by an abrupt wall deflection, Reinauer and Hager (1998) introduced the *shock diffractor*. The purpose of this pyramid-shaped element is to break up the compact shock wave and thus reduce its height. Figure 4.32 compares the shockwaves due to contraction without, and with shock diffractor and shows the positive effect in the tailwater chute.

The *optimum diffractor* shape was tested experimentally and found as shown in Fig. 4.33. The triangular shaped base of length $6h_o$ and

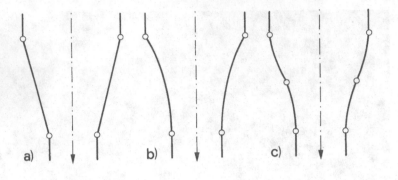

Figure 4.30
Contraction shapes
(a) funnel, (b) fan,
(c) nozzle

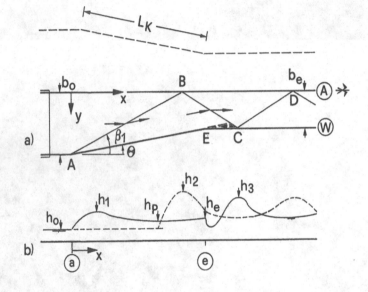

Figure 4.31
Chute contraction
(a) plan and
(b) side view with
(—) wall and (---)
axial surface profiles

width $4h_o$ increases to the point of abrupt wall deflection, where its height is $0.9h_o$, independent of the approach Froude number. The back of the element is vertical, and it may be aerated from the side to inhibit local cavitation damage of the diffractor. The element is small and may easily be added to existing chute contractions. A second diffractor may be added in the contraction axis to further reduce the tailwater waves. A thorough design is presented by Reinauer and Hager (1998).

The effect of a shock diffractor is the formation of a negative wave at its rear by which *wave diffraction* occurs, i.e. the shockwave is transversally stretched and thus reduced in height (Fig.4.32). To improve the contraction flow, a diffractor 1 is positioned with its vertical face at the contraction point A, and a diffractor 2 may be positioned in the contraction axis. The location of diffractor 2 is such that wave 1 is directed into its rear wave, and only a small wave 2 is generated. Off-design was found to have small consequences, and the diffractor should be designed for the flow with the largest shock number.

Figure 4.32
Effect of shock
diffractor at
contraction
(a) without, (b) with
diffractor element

(a)   (b)

For contraction flow with bottom elements, three waves may also be
identified:

- wall wave 1 downstream from contraction inlet,
- axis wave 2 at the centre of the contraction, and
- wall wave 3 downstream of the contraction end.

Without shockwave development, the increase of flow depth across a
contraction would be $h_3/h_1 = b_1/b_3$, for constant velocity. The *heights
of waves* 1 to 3 with shock diffractors thus are

$$Y_1 = h_1/h_o = (1 + 1.7S_o + 0.011F_0^2)/\cos\alpha, \qquad (4.54)$$

$$Y_2 = (1 + S_o)^2 \qquad (4.55)$$

$$Y_3 = \beta_e^{-1} - 0.2\alpha^{0.6} + 1.2S_o^{1/2}. \qquad (4.56)$$

The term in F_o in Eq.(4.54) stems from the self-perturbation of the
diffractor, and Y_1 with diffractor may be smaller or larger than without diffractor. Waves 2 and 3 are always much lower with a diffractor
than without the diffractor. The decision whether a diffractor should
be inserted depends mainly on an economical consideration involving

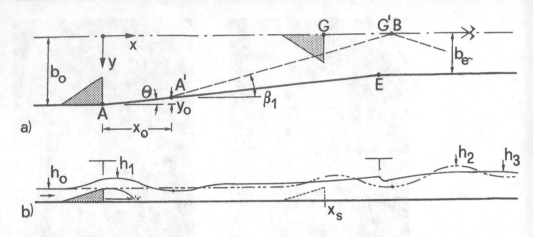

Figure 4.33 Definition sketch of shock diffractors in chute contraction a) plan, b) side view with reduced wave pattern. (—) Wall and (---) axial surface profiles

cost and reduction of wall height. Diffractors can easily be added to existing designs, and an aerator can be provided if cavitation damage is a concern. The chute contraction as presented here has its hydraulic optimum for width reductions between 50 to 80%. For $\beta_e < 0.5$ fan-shaped contractions should be considered (ICOLD, 1992).

Contractions with a supercritical approach flow are prone to *choking*, i.e. the breakdown of supercritical flow and the formation of a hydraulic jump. For such a contraction overtopping may occur

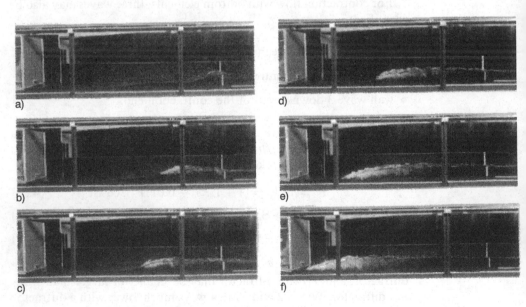

Figure 4.34 Choking phenomena in horizontal chute contraction with $\beta_e = 0.6$ for incipient choking at $F_o = 2.4$

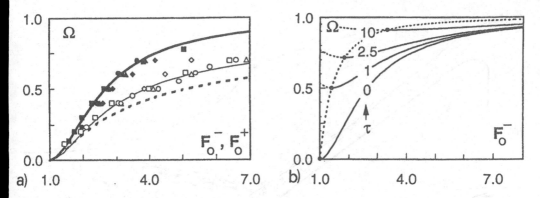

Figure 4.35 (a) Choking in *horizontal* contraction with $\Omega = 1 - \beta_e$, (—) incipient choking F_o^-, ($\cdots$) blowout for F_o^+ and (b) incipient choking in *sloping* contraction for various slope parameters τ. Note that choking occurs *above* a specific curve

because it was designed for a much smaller flow depth. A safety check against choking is required, therefore.

Choking flow was analysed by Henderson (1966). For *incipient choking* such as due to a reduction of discharge the flow at the end section of the contraction is critical. Applying the Bernoulli equation over the contraction relates the approach Froude number F_o to the width ratio β_e as

$$\beta_e = F_o \left(\frac{3}{2 + F_o^2} \right)^{3/2} \tag{4.57}$$

It was found from extended model observations that the choking phenomenon is stable in the sense that the toe of the hydraulic jump is shifted always slightly upstream of the contraction point A. Figure 4.34 shows a series of photographs illustrating sequences of choking.

For a choked contraction, a much larger approach Froude number is needed for *blowout*, i.e. for returning to supercritical contraction flow. Figure 4.35 refers to the *choking control*. In the horizontal channel one may distinguish between incipient choking F_o^- according to Eq.(4.57), and blowout for F_o^+. For sloping contractions, the parameter $\tau = (S_o - S_f)(L_k/h_o)$ controls F_o^- as shown in Figure 4.35(b), with S_o as bottom slope, S_f as average friction slope and L_k as the length of the contraction. For $0.5 < \beta_e < 1$, or $0 < \Omega < 0.5$ choking occurs only if $\tau > 1$. Choking is normally not a concern for chute slopes in excess of $5°$, therefore.

4.3 CASCADE SPILLWAY

Whereas the chute directs an overflow smoothly to the outlet structure, where a concentrated energy dissipation is applied, the cascade

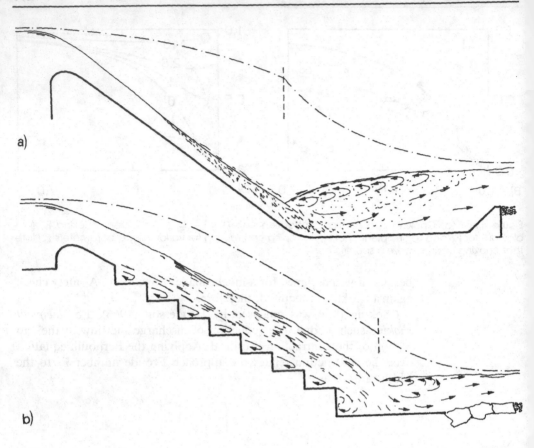

Figure 4.36 Comparison between (a) local and (b) continuous energy dissipation, (---) energy head line.

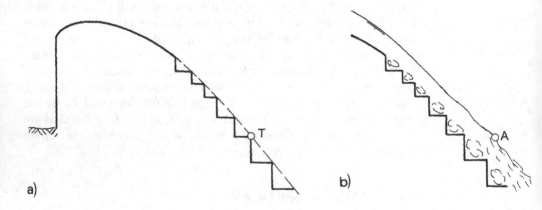

Figure 4.37 Stepped spillway (a) transition from crest to uniformly stepped spillway, (b) development of skimming flow and incipient air entrainment at point A

corresponds to a distributed dissipator. Accordingly, the terminal structure has a reduced rate of energy to dissipate, and is thus significantly smaller. Fig.4.36 compares the conventional system chute – stilling basin with the spillway cascade. The latter type is suited for small and medium discharges and has recently gained some popularity with RCC dams.

Figure 4.37(a) shows the geometry of the stepped spillway with the standard crest geometry (Chap. 2) and increasing step height up to the point of tangency T. Close to the crest the free surface is smooth although the development of a rolling vortex behind each step. The transition to rough surface flow is beyond point A where the incipient aeration occurs.

The *hydraulic features* of the cascade spillway as compared to chute flow are:

- the flow depth is much larger than in a chute due to the highly turbulent cascade flow, and higher sidewalls are required,

- much air is entrained and the spray action may become a concern,

- abrasion can be a serious problem for flows with sediment or with floating debris.

In cascade spillways two *flow types* may occur as shown in Figure 4.38 (Rajaratnam, 1990):

- *Nappe flow* where the flow from each step hits the step below as a falling jet;

- *Skimming flow* where the flow remains coherent over the individual steps. The onset of skimming flow occurs for $h_c/s > 0.8$, where h_c is the critical depth and s the step height. For long channels where uniform cascade flow may be attained, skimming flow dissipates more energy than nappe flow. However, nappe flow is more efficient for a short cascade than skimming flow (Chanson, 1994b).

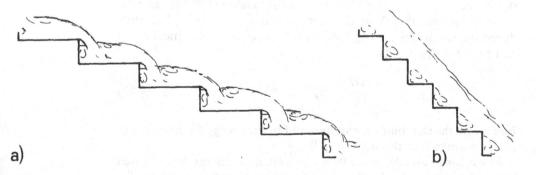

a)   b)

Figure 4.38 Cascade spillway with (a) nappe flow and (b) skimming flow

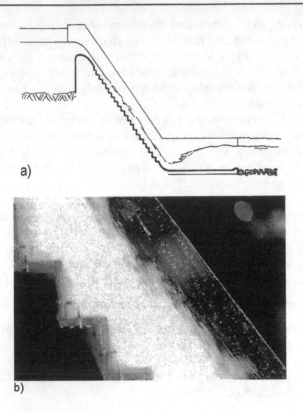

a)

b)

Figure 4.39
(a) Cascade
spillway with
tailwater stilling
basin, (b) cascade
flow (Diaz-Cascon,
et al., 1991)

The energy dissipated ΔH relative to the drop height H_o depends on the drop Froude number $\mathbf{F}_o = q/(gH_o^3)^{1/2}$ and the slope Θ of the spillway. With Θ in deg., Stephenson (1991) found

$$\frac{\Delta H}{H_o} = \left(\frac{0.84}{\Theta^{0.25}}\right)\mathbf{F}_o^{-1/3}. \tag{4.58}$$

Accordingly, the effect of slope is small, whereas the dam height has a significant influence on the head loss. Christodoulou (1993) studied the effect of step number N on the energy dissipation $\Delta H/H_o$. He introduced the parameter $y_c = h_c/(Ns)$ with $h_c = (q^2/g)^{1/3}$ as critical depth and found for $y_c < 0.25$

$$\frac{\Delta H}{H_o} = \exp(-30y_c^2). \tag{4.58}$$

Increasing the step number can thus add to the energy dissipation and the performance of the stepped spillway.

For a long cascade, some 90% of mechanical energy is dissipated along the cascade, and only a small portion of energy must be dissipated in the stilling basin. According to Stephenson (1991) the *effi-*

ciency of the cascade spillway depends mainly on its height and the specific discharge and only slightly on the slope. The cascade flow may reach a state of nearly *uniform flow* (subscript u) which can be approximated with L_s as step length (Vischer and Hager, 1995)

$$h_u = 0.23[L_s^4 q^6/(sg^3)]^{1/12}. \tag{4.59}$$

Diez-Cascon, et al. (1991) tested a cascade spillway of step size $L_s/s = 0.75$ and slope $\Theta = 53°$, followed by a horizontal stilling basin (Figure 4.39). The *sequent depth ratio* $Y = h_2/h_1$ varied with the approach Froude number $\mathbf{F}_1$ as

$$Y = 2.9\mathbf{F}_1^{2/3}. \tag{4.60}$$

The resulting tailwater depth is larger than for the classical hydraulic jump (Chapter 5), but the value $\mathbf{F}_1$ for cascade flow is usually much smaller. According to Eq.(4.59) valid for uniform approach flow one deduces $\mathbf{F}_u^{2/3} = 7.3(h_u/L_s)^{1/3}$ and $Y = 21.1(h_u/L_s)^{1/3}$ instead of Eq.(4.60). The sequent depth ratio varies slightly with the uniform flow depth relative to the step height.

According to Chanson (1994a) the *onset of nappe flow* occurs for $h_c > h_c^*$ where

$$\frac{h_c^*}{s} = 1.057 - 0.465\frac{s}{L_s}. \tag{4.61}$$

In the transitional regime between nappe and skimming flow a *hydraulic instability* was reported which should be avoided to inhibit problems with structural vibration. For skimming flow, the resistance characteristics are governed by the distance of two adjacent step edges, protruding into the flow. Although Chanson (1994a) has analysed the hydraulics of skimming flow, there is currently a lack of data to describe uniform cascade flow. A fundamental research study is needed to obtain further information.

REFERENCES

Chanson, H. (1989a). Study of air entrainment and aeration devices. *Journal Hydraulic Research* **27**(3): 301–319.

Chanson, H. (1989b). Flow downstream of an aerator – aerator spacing. *Journal Hydraulic Research* **27**(4): 519–536.

Chanson, H. (1994a). Hydraulics of skimming flows over stepped channels and spillways. *Journal Hydraulic Research* **32**(3): 445–460.

Chanson, H. (1994b). State of the art of the hydraulic design of stepped chute spillways. *Journal Hydropower and Dams* **1**(4): 33–42.

Chow, V.T. (1959). *Open channel hydraulics*. McGraw Hill, New York.

Christodoulou, G.C. (1993). Energy dissipated of stepped spillways. *Journal Hydraulic Engineering* **119**(5): 644–650.

Dias, F. and Tuck, E.O. (1991). Weir flows and water falls. *Journal Fluid Mechanics* **230**: 525–539.

Diez-Cascon, J., Blanco, J.L., Revilla, J., Garcia, R. (1991). Studies on the hydraulic behaviour of stepped spillways. *Water Power & Dam Construction* **43**(9): 22–26.

Ervine, D.A., Falvey, H.T., Kahn, A.R. (1995). Turbulent flow structure and air uptake at aerators. *Journal Hydropower and Dams* **2**(4): 89–96.

Falvey, H.T. (1990). Cavitation in chutes and spillways. *Engineering Monograph* **42**. Water Resources Technical Publication. US Printing Office, Bureau of Reclamation, Denver.

Hager, W.H. (1991). Uniform aerated chute flow. *Journal Hydraulic Engineering* **117**(4): 528–533.

Hager, W.H. and Mazumder, S.K. (1992). Supercritical flows at abrupt expansions. Proc. Institution Civil Engineers *Water Maritime and Energy* **96**: 153–166.

Henderson, F.M. (1966). *Open channel flow*. MacMillan, New York.

ICOLD (1992). Spillways. Shockwaves and air entrainment. *Bulletin* **81**, prepared by W.H. Hager. International Commission of Large Dams, Paris.

Ippen, A.T. and Dawson, J.H. (1951). Design of channel contractions. *Trans. ASCE* **116**: 326–346.

Jansen, R.B. (1988). *Advanced dam engineering for design, construction and rehabilitation*. Van Nostrand Reinhold, New York.

Kells, J.A. and Smith, C.D. (1991). Reduction of cavitation on spillways by induced air entrainment. *Canadian Journal Civil Engineering* **18**: 358–377; **19**: 924–929.

Knapp, R.T. (1951). Design of channel curves for supercritical flow. *Trans. ASCE* **116**: 296–325.

Koolgaard, E.B. and Chadwick, W.L. (1988). *Development of dam engineering in the United States*. Pergamon Press, New York.

Mazumder, S.K. and Hager, W.H. (1993). Supercritical expansion flow in Rouse modified and reversed transitions. *Journal Hydraulic Engineering* **119**(2): 201–219.

Naudascher, E. and Rockwell, D. (1994). Flow-induced vibrations. *IAHR Hydraulic Structures Design Manual* **7**. Balkema: Rotterdam, Brookfield.

Rajaratnam, N. (1990). Skimming flow in stepped spillways. *Journal Hydraulic Engineering* **116**(4): 587–591; **118**(1): 111–113.

Rajaratnam, N., Subramanya, K., Muralidhar, D. (1968). Flow profiles over sharp-crested weirs. *Journal Hydraulics Division* ASCE **94**(HY3): 843–847.

Reinauer, R. and Hager, W.H. (1996). Generalized drawdown curve for chutes. Proc. Institution Civil Engineers, *Water Maritime and Energy* **118**(4): 196–198.

Reinauer, R. and Hager, W.H. (1997). Supercritical bend flow. *Journal Hydraulic Engineering* **123**(3): 208–218.

Reinauer, R. and Hager, W.H. (1998). Supercritical flow in chute contraction. *Journal Hydraulic Engineering*, **124**(1): 55–64.

Rouse, H., Bootha, B.V., Hsu, E.Y. (1951). Design of channel expansions. *Trans. ASCE* **116**: 1369–1385.

Rutschmann, P. and Hager, W.H. (1990a). Air entrainment by spillway aerators. *Journal Hydraulic Engineering* **116**(6): 765–782; **117**(4): 545; **118**(1): 114–117.

Rutschmann, P. and Hager, W.H. (1990b). Design and performance of spillway chute aerators. *Water Power & Dam Construction* **41**(1): 36–42.

Schwalt, M. and Hager, W.H. (1992). Shock pattern at abrupt wall deflection. ASCE Conf. Environmental Engineering *Water Forum '92* Baltimore: 231–236.

Schwartz, H.I. (1964). Projected nappes subject to harmonic pressures. *Proc. Institution Civil Engineers* **28**: 313–325.

Stephenson, D. (1991). Energy dissipation down stepped spillways. *Water Power & Dam Construction* **43**(9): 27–30.

Straub, L.G. and Anderson, A.G. (1960). Experiments on self-aerated flow in open channels. *Trans. ASCE* **125**: 456–486.

Thomas, H.H. (1976). *The engineering of large dams.* John Wiley & Sons: London, New York.

Toth, I. (1968). Le barrage Gardiner sur la rivière South Saskatchewan, Canada. *Technique des Travaux* **44**: 363–374.

USBR (1948). Studies of crests for overfall dams. Boulder Canyon Projects – Final Reports. Part VI – Hydraulic Investigations. *Bulletin* **3**. US Bureau of Reclamation, Dept. Interior, Denver.

Vanden-Broeck, J.M. and Keller, J.B. (1986). Pouring flows. *Physics Fluids* **29**(12): 3958–3961.

Vischer, D.L. (1988). A design principle to avoid shockwaves in chutes. Int. Symp. *Hydraulics for High Dams* Beijing: 391–396.

Vischer, D.L., Volkart, P., Siegenthaler, A. (1982). Hydraulic modelling of air slots in open chute spillways. *Hydraulic Modelling of Civil Engineering Structures* Coventry: 239–252.

Vischer, D.L. and Hager, W.H. (1995). Energy dissipators. *IAHR Hydraulic Structures Design Manual* **9**. Balkema: Rotterdam, Brookfield.

Volkart, P. and Rutschmann, P. (1984). Air entrainment devices. *Mitteilung* **72**. Versuchsanstalt für Wasserbau, Hydrologie und Glaziologie, ed. D. Vischer. Swiss Federal Institute of Technology, Zurich.

Wood, I.R. (1985). Air water flows. *XXI IAHR Congress* Melbourne **6**: 18–29.

Wood, I.R. (1991). Air entrainment in free-surface flows. *IAHR Hydraulic Structures Design Manual* **4**. Balkema: Rotterdam, Brookfield.

Wood, I.R., Ackers, P., Loveless, J. (1983). General method for critical point on spillways. *Journal Hydraulic Engineering* **109**(2): 308–312.

Stilling basin of Gardiner dam, Saskatchewan Canada (*Gardiner dam* Prairie Farm Rehabilitation Administration, 1980)

5

Dissipation Structures

5.1 INTRODUCTION

The hydraulic jump is the basic feature used to dissipate excess hydro-mechanical energy. It is the most discontinuous and turbulent flow in an open channel. The prominent features of the hydraulic jump are:

- highly turbulent flow,

- pulsations in the body of jump,

- air entrainment at the toe and air detrainment at the end of jump,

- generation of spray and sound,

- energy dissipation due to turbulence production, and

- erosive potential and generation of tailwater waves.

A hydraulic jump may occur under two different ways: (1) as a transition from supercritical to subcritical flow in a channel with a varied location, or (2) in a stilling basin with a fixed location.
Hydraulic jumps may have various appearances, such as the:

- *classical hydraulic jump* as the basic jump type occurring in a prismatic, horizontal and quasi-frictionless rectangular channel,

- *sloping hydraulic jump* as a hydraulic jump in a sloping, normally rectangular channel,

- *submerged hydraulic jump* when the toe of the jump is covered by the tailwater, and

- *expanding hydraulic jump* in a nearly horizontal channel with a diverging cross-section.

The main features of a hydraulic jump are its effectiveness for energy dissipation and its stability under varying boundary conditions. Because jumps with a non-rectangular cross-section are prone to instability, they are rarely used in dam hydraulics. For a review of such jumps, reference is made to Elevatorsky (1959), Rajaratnam (1967), McCorquodale (1986), and Hager (1992).

The *stilling basin* uses the hydraulic jump as the hydraulic element for dissipating energy in a specific structure. A distinction between the hydraulic jump basin and the baffle basin can be made. In the *hydraulic jump basin* a structure with a nearly horizontal and smooth bottom is used to fix the location of dissipation. The basin is designed to be resistant against spray, scour, pulsations, cavitation, tailwater waves and blowout (Figure 5.1). Such basins are recommended for small approach energy heads less than 10m, or medium heads between 30 and 50 m (Mason, 1982). For larger heads, cavitation becomes a problem and the basin is no longer safe against cavitation damage.

For small to medium approach energy heads of 10 to 30 m, a stilling basin provided with baffles can be used to shorten the structure. Elements such as blocks, steps or sills are located at the basin bottom by which the approach flow is deflected to the surface, yet without inducing excessive pulsations.

A general account for both the hydraulic jump and hydraulic jump stilling basins is given in Section 5.2. For the latter, some standardized designs are presented and it is recommended that these be considered before a novel design is introduced. Actually, the number of basins introduced is so large that novel designs can only be justified if particular site conditions prevail. For larger basins, extended model observations are recommended because of the unexpected features

Figure 5.1 Stilling basin of Manitoba Hydro Limestone (Canada) with one turbine in operation (*Water Power and Dam Construction*, July 1991)

Figure 5.2
Drop structures of
Wagendrift dam,
South Africa (*Journal
of Hydropower and
Dams*, November
1994)

that can occur even for small deviations from tested designs. *Plunge pools* of drop structures in general are intermediate to hydraulic jumps and trajectory basins, and both elements play a role. Its main features are described in Chapter 4 and in Section 5.3. Figure 5.2 refers to typical drop structures as applied with arch dams.

For heads larger than, say 50 m, the *trajectory basin* is a standard design as described in 5.4. For such basins the effect of jet dispersion is used to dissipate a portion of the excess energy (Fig.5.3). The knowledge on trajectory basins is limited when compared to the hydraulic jump stilling basins and aspects of scour need to be carefully investigated.

a)

b)

Figure 5.3 Trajectory basins of Tarbela dam (Pakistan) (a) auxiliary spillway (*ICOLD* Q69, R24), (b) outlet structure.

5.2 HYDRAULIC JUMP AND STILLING BASIN

5.2.1 Classical hydraulic jump

The basic flow type of a hydraulic jump is referred to as classical. As mentioned above, it occurs in a straight prismatic horizontal channel of rectangular shape in which wall friction is negligible. All notation referring to the classical hydraulic jump are denoted with an asterisk. Major contributions towards the classical hydraulic jump were made by Rouse, et al. (1959) in presenting the turbulence characteristics, by Schröder (1963) and Rajaratnam (1965) in relating it to the *wall jet* phenomenon and by McCorquodale and Khalifa (1983) in presenting a numerical approach.

Figure 5.4 shows a schematic of the classical hydraulic jump, or in short the classical jump. Its body is located between the toe and the end of the jump. Just upstream from the toe, the flow depth is h_1, the average velocity is V_1, and x is the longitudinal coordinate. At the end of the jump, the depth of flow is h_2^*, the velocity is V_2^* and the length of the jump is L_j^*. All quantities are time-averaged because of the significant pulsating action. The approach flow expands along the bottom of the body as a wall jet and rises to the surface where a stagnation point forms at location L_r^* downstream of the toe. A portion of the

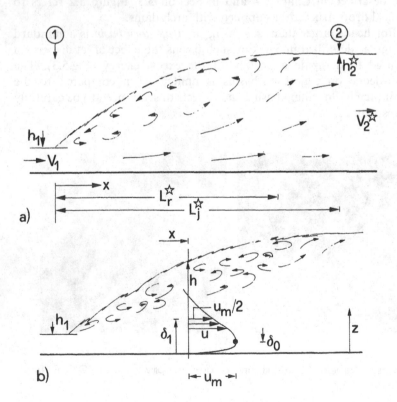

Figure 5.4
Classical hydraulic jump (a) length and (b) velocity characteristics.

flow returns as a *surface roller* that is entrained in the toe zone, whereas the remainder is deflected in the tailwater. Due to the large velocity gradient between the wall jet and the roller flow, air is entrained close to the toe section, and the turbulence level is significant.

The classical hydraulic jump is an example where the basic momentum equation can be successfully applied, and where the energy loss across the jump may be predicted. The following discussion refers to the gross flow pattern first, and the internal flow features associated with a classical jump afterwards.

The ratio of *sequent depths* $Y^* = h_2^*/h_1$ may be computed when assuming in sections 1 and 2:

- hydrostatic pressure distributions,

- uniform velocity distributions,

- air entrainment is negligible, and

- time-averaged quantities.

The momentum equation reads with ρ as density, g as gravitational acceleration and b as channel width (Figure 5.4(a))

$$\frac{1}{2}\rho g b h_1^2 + \rho Q V_1 = \frac{1}{2}\rho g b h_2^{*2} + \rho Q V_2^*. \tag{5.1}$$

The discharge is $Q = b h_1 V_1 = b h_2^* V_2^*$. With the *approach Froude number*

$$\mathbf{F}_1 = \frac{Q}{(g b^2 h_1^3)^{1/2}} \tag{5.2}$$

the solution of Eq.(5.1) is

$$Y_* = \frac{1}{2}[(1 + 8\mathbf{F}_1^2)^{1/2} - 1]. \tag{5.3}$$

For Froude numbers larger than $\mathbf{F}_1 = 2$ as are considered subsequently, this approximates as

$$Y^* = \sqrt{2}\mathbf{F}_1 - (1/2). \tag{5.4}$$

For a given depth h_1, the sequent depth h_2^* increases thus linearly with $\mathbf{F}_1$.

The *efficiency* $\eta^* = (H_1 - H_2^*)/H_1$ of the classical hydraulic jump with H as the energy head can be determined with the energy equation to yield, again for $\mathbf{F}_1 > 2$

$$\eta^* = \left(1 - \frac{\sqrt{2}}{\mathbf{F}_1}\right)^2.$$

For $F_1 < 3$ the relative loss of head is thus less than 30%, for $F_1 = 5$ it is 50% and η^* is over 70% for $F_1 > 9$.

Classical jumps have various appearances, depending on the approach Froude number. For $F_1 < 1.6$ to 1.7 jumps are referred to as undular because of the undulating surface pattern. They are inefficient for energy dissipation and generate tailwater waves (Reinauer and Hager, 1995). For $1.7 < F_1 < 2.5$, the efficiency is still small and (a) pre-jumps are generated. The (b) transition jump with $2.5 < F_1 < 4.5$ has a pulsating action. Such jumps occur often behind low head structures. The best performance has the (c) stabilized jump where $4.5 < F_1 < 9$. It is compact and stable with a considerable dissipation and a good stilling action. The (d) choppy jump for $F_1 > 9$ is rough and overforced (Figure 5.5).

The *length of roller* is approximately

$$L_r^*/h_2^* = 4.5 \qquad\qquad (5.6)$$

and the *length of jump* is nearly equal to

$$L_j^*/h_2^* = 6. \qquad\qquad (5.7)$$

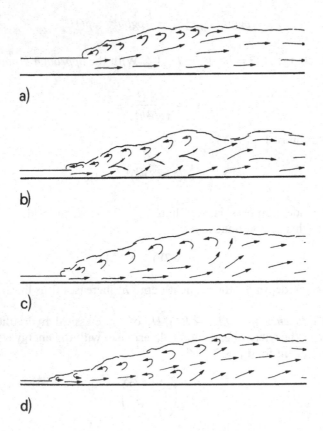

Figure 5.5
Forms of classical
hydraulic jump
(Peterka, 1958)

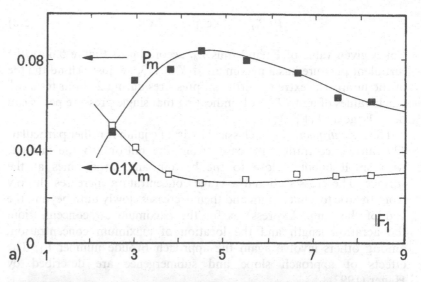

a)

b)

Figure 5.6
(a) Maximum rms-pressure fluctuation P_m and corresponding location X_m as functions of F_1 (Vischer and Hager, 1995), (b) Damages on stilling basin (*Water Power & Dam Construction* 39(5): 56)

Vischer and Hager (1995) have also presented detailed information on the internal flow pattern of the classical hydraulic jump. All quantities depend exclusively on the approach Froude number F_1 and on the approach flow depth h_1.

The *turbulent pressure characteristics* are important in designing the cavitation resistance. With p' as the fluctuating pressure component and p_f as root-mean-square (rms)-value the dimensionless pressure value $P = p_f/(\rho V_1^2/2)$ varies only with the dimensionless location $X = x/L_r^*$. The maximum P_m of the function $P(X)$ varies also with F_1 according to Figure 5.6(a) and is 0.08 for $F_1 = 4.5$, and smaller for other F_1. The distribution of pressure $P(X)$ along the jump can be expressed as

$$P/P_m = [3X\exp(1 - 3X)]^2. \qquad (5.8)$$

For a given value of $\mathbf{F}_1$ and thus P_m according to Figure 5.6(a), the turbulent pressure has a maximum at $X = 1/3$, i.e. just behind the toe of the jump. The extreme turbulent pressures during 24 hour tests can reach values of up to $P_m = 1$, indicating that single pressure peaks can be as large as $\pm(V_1^2/2g)$.

The *air entrainment* in a classical hydraulic jump is rather particular. The air concentration increases from the bottom to the surface, with small bubbles close to the bottom and larger ones at the surface. The cross-sectional average concentration increases sharply from the toe to a maximum and then decreases slowly until beyond the end of the jump. Expressions for the maximum air concentration, the aeration length and the location of maximum concentration, among others involve again the approach Froude number $\mathbf{F}_1$. The effects of approach slope and submergence are described by Hager (1992).

5.2.2 Stilling basins

In a stilling basin excess hydromechanical energy is converted mainly into heat, spray and sound. The stilling basin is a hydraulic structure located between the outlet works of a dam and the tailwater, to where it should return excess flows safely. The stilling basin is a structure in which a hydraulic jump is generated and which has been designed economically in terms of length, tailwater level and scour.

The *selection* of a stilling basin depends on factors such as:

• hydraulic approach conditions,

• tailwater characteristics,

• scour potential, and

• personal preferences.

The approach energy head should be between 10 and 30 m, in order that the performance of the basin is successful. Actually, a number of standard basins are available that have been tested extensively (Peterka, 1958). Problems with stilling basins can occur for high approach velocity, Froude number smaller than 2.5, asymmetric approach conditions, curved in- or outflow, or low tailwater level. In general, the minimum tailwater level should be equal to the sequent depth as given in Eq.(5.4). Baffles are provided mainly to shorten the hydraulic jump without gaining additional tailwater level.

Compared to a simple hydraulic jump basin, in which the approach momentum is balanced by an adequate tailwater level, the stilling basin

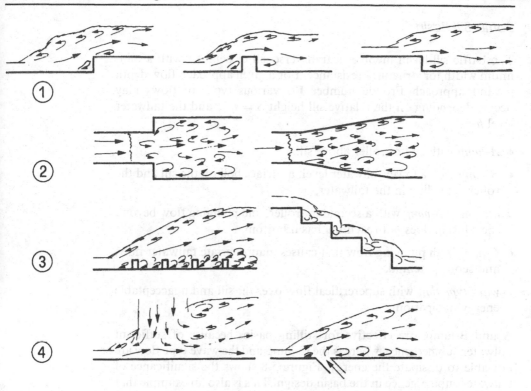

Figure 5.7 Basic elements of stilling basins (Hager, 1992). ① Bottom geometry, ② geometry of plan, ③ boundary roughness, ④ discharge addition

has in addition chute and baffle elements. Those elements are located on the basin bottom and involve steps, sills or blocks. The effect of dissipation can be increased with a diverging basin. Increasing the wall roughness and adding lateral discharge have not received attention in practice, mainly because of problems with cavitation and stability. Figure 5.7 shows a number of elements that have been discussed as possibilities to increase one or the other aspect of stilling basin flow.

Among the large variety of shapes that have been proposed (Vischer and Hager, 1995) the baffle sill basin and the baffle block basin are certainly the most popular designs. Those elements are prone to approach heads larger than 30 m, as previously mentioned, due to *cavitation damage. Abrasion* is not a concern for basins downstream of overflow structures but for bottom outlets it may be a serious problem. For stilling basins subjected to flows with a high abrasion potential, baffles should have a steel-armouring, or a simple basin should be provided. In the following baffle basins together with the abruptly expanding basin are discussed. The USBR standard basins as introduced by Peterka (1958) are not redescribed here.

Baffle Sill Basin

The baffle sill basin involves a transverse sill of height s with a minimum width for structural resistance. For a given approach flow depth h_1 and approach Froude number $\mathbf{F}_1$, various types of flows may occur, depending on the relative sill height $S = s/h_1$ and the tailwater level h_2:

- *A-jump* with end of roller above sill,

- *B-jump* with a lower tailwater level, a surface boil on the sill and the roller extending in the tailwater,

- *minimum B-jump* with a secondary roller, and plunging flow beyond the sill that does not reach the basin bottom,

- *C-jump* with plunging flow that causes inappropriate tailwater flow, and scour potential,

- *wave type flow* with supercritical flow over the sill and unacceptable energy dissipation.

A and B-jumps are effective for stilling basins because of sufficient tailwater submergence, whereas the C-jump and the wave type flow are not able to dissipate the energy. Figure 5.8 shows the significance of tailwater submergence in the basin design. This is also an example that illustrates that a slight decrease of tailwater level below the sequent depth as obtained with Eq.(5.4) has dramatic consequences on the performance of a stilling basin. The purpose of any baffle element should thus involve a *length reduction* but no significant reduction of the tailwater level required.

Figure 5.9 compares the baffle sill basin and the hydraulic jump basin. The sill is defined with the relative height $S = s/h_1$ and the relative sill location $\Lambda = L_s/L_r^*$. The *sequent depth* ratio required $Y = Y^* - \Delta Y_s$ is composed of the effect of classical jump and the effect of sill (Hager, 1992)

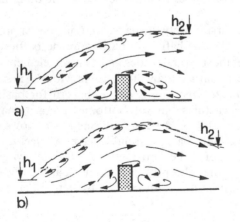

a)

b)

Figure 5.8
Stilling basin with (a) sufficient submergence and effective dissipation, (b) insufficient tailwater submergence

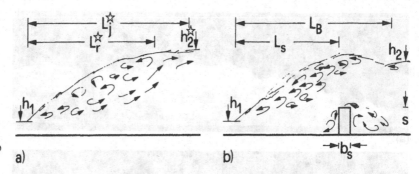

Figure 5.9
Comparison of (a) classical hydraulic jump and (b) baffle sill basin

a)   b)

$$\Delta Y_s = 0.7S^{0.7} + 3S(1 - \Lambda)^2. \tag{5.9}$$

For any sill height S, a *minimum* approach Froude number F_{1m} is necessary for the formation of the hydraulic jump, and the corresponding *maximum* sill height S_M for any approach Froude number is

$$S_M = \frac{1}{6}F_1^{5/3}. \tag{5.10}$$

The relative sill height should be limited in practice to $S_M = 2$. Also, the sill should neither be too small nor too large to inhibit ineffectiveness and overforcing. The *optimum* sill height S_{opt} is

$$S_{\text{opt}} = 1 + \frac{1}{200}F_1^{2.5}. \tag{5.11}$$

Depending mainly on the relative sill position Λ, three types of jump may appear:

1. $\Lambda > 0.8$ (to 1) *A-jump* with practically no scour potential and suitable for easily erodible beds,

2. $0.65 < \Lambda < 0.8$ *B-jump* with small erosion mainly along the tailwater walls,

3. $0.55 < \Lambda < 0.65$ *minimum B-jump* only suitable for rocky tailwater channels.

The *length of the jump* L_j from the toe to the end of the bottom roller relative to the length of the classical jump L_j^* is

$$L_j/L_j^* = 1 - 0.6S^{1/3}(1 - \Lambda). \tag{5.12}$$

The *length of basin* L_B (Figure 5.9(b)) under all three types of flows is slightly less than the length of a classical jump. A sill basin improves the *stabilization* of a hydraulic jump under variable tailwater and is somewhat shorter than a classical hydraulic jump. The effect of

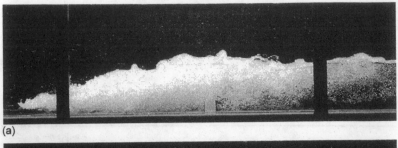

(a)

(b)

Figure 5.10.
Photographs of (a)
A-jump and (b) B-
jump for $F_1 = 5$

additional appurtenances is discussed below. Figure 5.10 shows photographs of the A- and B-jumps in a lab channel.

Baffle Block Basin

As for the baffle sill basin, a baffle block basin involves various types of flows. For optimum basin flow, the blocks must have an adequate location and height to counter ineffective or overforced flow. According to Basco (1971) the optimum height $S_{opt} = s_{opt}/h_1$ and the optimum basin length are, respectively

$$S_{opt} = 1 + \frac{1}{40}(F_1 - 2)^2, \tag{5.13}$$

$$(L_B/h)_{opt} = 1.6 + 7.5F_1^{-2}. \tag{5.14}$$

Figure 5.11 shows the basin with the standard USBR blocks, where the block spacing is equal to the block width $e = b_B$ and $e/s = 0.75$. The force on the blocks F_B may be described with the *force coefficient* $\Phi = F_B/[\rho g b h_2^{*2}/2]$ where Φ for optimum basin performance is

$$\Phi_{opt} = \frac{1}{7} + \frac{F_1}{100} \tag{5.15}$$

and the *sequent depth ratio* obtains

$$Y = \left(\frac{2}{1+\Phi}\right)^{1/2} F_1 - \frac{1}{2}. \tag{5.16}$$

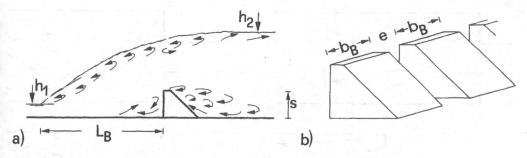

Figure 5.11 Baffle block basin (a) side view, (b) standard block geometry

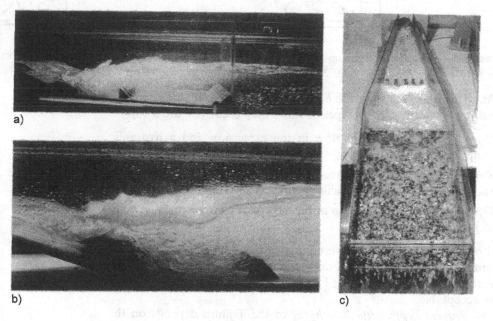

Figure 5.12 Stilling basin with chute blocks, baffle blocks and end sill at design discharge

The tailwater reduction is thus over 10% as compared with the classical jump. A staggered block row was insignificant in the performance of the basin. Details on the pressure characteristics are given by Hager (1992). Figure 5.12 refers to a typical stilling basin tested in the laboratory.

Expanding Stilling Basin

Stilling basins can expand behind outlets with partial operation or because of width increase from the approach channel to the tailwater. In practice, the abrupt expansion is of interest due to structural compactness. In the following this geometry is considered and reference is made to Hager (1992) for gradually expanding basins.

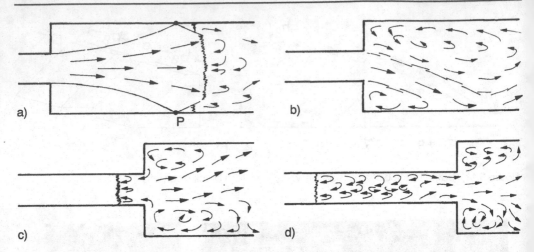

Figure 5.13 Types of flow in abruptly expanding stilling basin without bottom elements (a) R-jump, (b) S-jump, (c) T-jump, (d) classical jump in approach channel

At an abrupt expansion with h_1, b_1, $\mathbf{F}_1$ as approach conditions and b_2 as tailwater width, various types of flows may occur (Figure 5.13):

- *R-jump* with supercritical flow in the expansion and a hydraulic jump in the tailwater reach,
- *S-jump* with the toe of jump more upstream but still in the expansion and the generation of an oscillating or even asymmetric jet flow,
- *T-jump* with the toe of jump in the approach channel and the body of jump in the expansion.

Although the R-jump is a stable configuration, it may develop into the S-jump for slightly increasing tailwater. The S-jump is highly spatial, unstable, and excessively long. For an expanding basin, the T-jump is only acceptable.

The *sequent depth ratio* $Y = h_2/h_1$ of the T-jump depends on the relative toe location $X_1 = x_1/L_r^*$ with L_r^* as roller length of the classical jump, the width ratio $\beta = b_2/b_1$ and the sequent depth ratio Y^* of the classical jump (Figure 5.14(a)). According to Hager (1992)

$$\frac{Y^* - Y}{Y^* - 1} = (1 - \beta^{-1/2})[1 - \tanh(1.9X_1)]. \qquad (5.17)$$

For $\beta = 1$ the asymptotic result is $Y = Y^*$. Also for $X_1 > 1.3$, the toe of jump is located so much in the approach channel that the end of jump is upstream from the expansion section, and $Y = Y^*$ (Figure 5.13(d)).

The efficiency of the T-jump increases as X_1 decreases, but the performance decreases in parallel so that the T-jump is not an effective dissipator. The performance of expanding stilling basins can be improved with a *baffle sill*. The following comments refer also to basins with an expansion angle larger than 30° (Bremen and Hager, 1994).

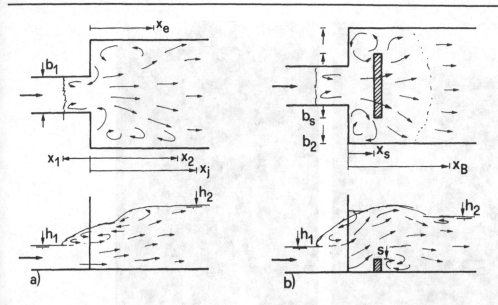

Figure 5.14 Abruptly expanding stilling basin (a) without and (b) with central sill. Plan (top) and section (bottom).

The optimized *expanding stilling basin* (Figure 5.14(b)) involves a transverse central sill of width

$$b_s/b_1 = 1 + (1/4)(\beta - 1). \qquad (5.18)$$

The sill forces the approach jet to expand and induces two *corner vortices* (Figure 5.15). The sill has a relative position $X_s = x_s/L_r^*$ and a relative height $S = s/h_1$ that are correlated to the approach Froude number $\mathbf{F}_1 = Q/(gb_1^2h_1^3)^{1/2}$ as

$$X_s = \frac{3}{4}(S/\mathbf{F}_1)^{3/4}. \qquad (5.19)$$

The sill height increases with increasing relative position, and approach Froude number, therefore. The sill position should be contained within

$$0.4 < \frac{x_s/b_1}{(\beta - 1)} < 0.6, \qquad (5.20)$$

and a reasonable sill height is

$$S < (1/2)(\mathbf{F}_1 - 1). \qquad (5.21)$$

The basin length x_B is equal to the length of roller L_r^*, when a conventional end sill is added. Conditions to be satisfied include (1) $3 < \mathbf{F}_1 < 10$ for the approach flow, (2) $1 < \beta < 5$ for the width ratio, and (3) $0.1 \le X_s \le 0.6$ for the relative sill position, next to those introduced for stilling basins in general. Figure 5.15 shows the performance of the expanding stilling basin in a lab channel.

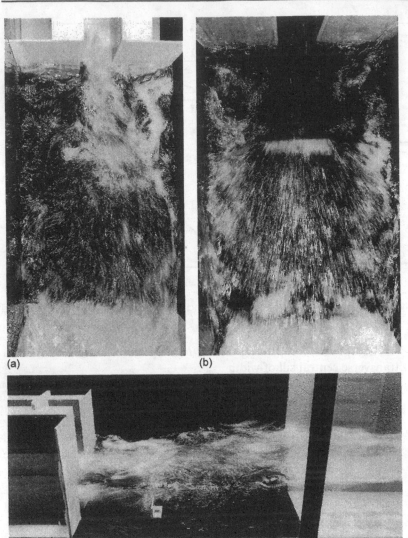

(a)   (b)

(c)

Figure 5.15
Expanding stilling
basin (a) without
and (b) with corner
vorticity well
developed, (c) side
view ($F_1 = 5, \beta = 5$)

Slotted Bucket Stilling Basin

Beichley and Peterka (1959) developed the slotted bucket stilling basin
with an 8° sloping apron on which teeth of 45° are mounted (Figure
5.16). The teeth involve stable flow and little boil action.

Three *types of flows* may be distinguished:

- sweep out with a tailwater level too low,

- *minimum* tailwater level below which excessive surface waves and
  scour occur, and

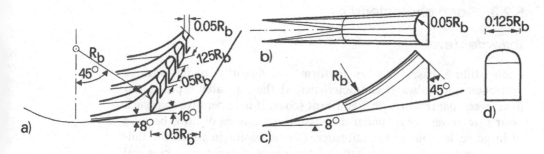

Figure 5.16 Slotted bucket stilling basin, notation and geometry

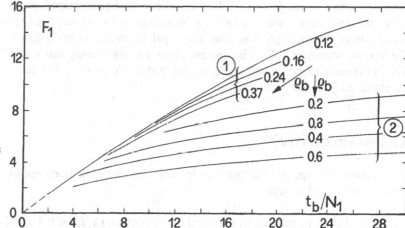

Figure 5.17
Extreme tailwater
levels for *slotted*
stilling basin t_b/N_1 as
function of $\mathbf{F}_1$ and
$\rho_b = (R_b/N_1)$
$[1 + (1/2)\mathbf{F}_1^2]$. ①
minimum and ②
maximum tailwater
depths

- *maximum* tailwater level above which dividing flow results.

A slotted bucket basin has a lower and an upper *limit of operation*. These depend on the approach Froude number $\mathbf{F}_1 = V_1/(gN_1)^{1/2}$ and the relative bucket radius $\rho_b = (R_b/N_1)[1 + (1/2)\mathbf{F}_1^2]$ where V_1 is the approach velocity, N_1 the flow depth measured perpendicular to the chute bottom, and R_b the bucket radius (Figure 5.16(c)).

The *minimum* bucket radius should be $R_{bm}/N_1 = 2.2\mathbf{F}_1^{1/2}$ and the extreme tailwater levels t_b/N_1 are given in Figure 5.17 as a function of ρ_b and $\mathbf{F}_1$. Care should be taken against material entering the bucket that may cause damage by *abrasion*. Figure 5.18 refers to a basin with optimum roller action.

Figure 5.18
Slotted bucket
stilling basin in lab
model (*L'Acqua* 1970
48(5): 125)

5.2.3   Basin characteristics

Tailwater Level

Each stilling basin has to perform satisfactorily under various approach and tailwater characteristics. If the tailwater depth is too low, *sweep-out* results with significant scour. It is imperative that such poor flow never occurs under all possible conditions of flow, because of large scale scour in the tailwater. For all basins in which the tailwater depth is lower than 90% of the sequent depth of a classical hydraulic jump, the design should be verified in sufficiently large model basins. Damage of a stilling basin has to be countered under any circumstances since it may have implications to the safety of the entire dam. Accordingly, the *tailwater depth-discharge relation* has to be known whenever stilling basins are chosen as the energy dissipator. Other aspects to be satisfied are the cavitation resistance, the scour control, and the tailwater waves.

Cavitation Control

Cavitation damage may occur due to *turbulent pressure fluctuations*. Two cases are of concern:

1. zone of peak turbulence in the front basin portion $(x/L_r^* < 0.4)$, and

2. zone of appurtenances, i.e. at the rear of baffle sills and baffle blocks.

The *lining* of a stilling basin is related to the pressure fluctuations, and a minimum slab thickness should be $0.27(V_1^2/2g)$. Two design conditions have to be met (Hager, 1992):

1. design discharge including the unbalanced uplift force, and

2. maximum reservoir level with empty stilling basin.

Scour Control

The transition from a stilling basin to the unprotected tailwater is important relative to the scour. The scour potential may greatly be reduced by providing suitable *end sills* that deflect remaining bottom currents to the surface and induce a bottom return current (Figure 5.19). According to Novak (1955) and denoting by z_t the tailwater bed elevation relative to the basin, and by H_1 the height of the dam above the stilling basin:

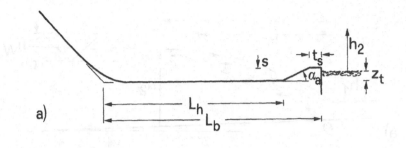

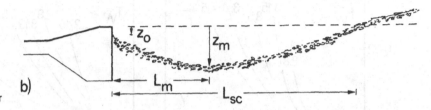

Figure 5.19
Scour control for
stilling basins (a)
basin geometry,
(b) tailwater scour

- the *apron angle* $\alpha_a = 20°$ minimizes the scour area and the location of maximum scour,

- the *sill width* t_s should have the structural minimum,

- the *sill height* s should satisfy the conditions $z_t/s \geq 0.4$ and $(s - z_t)/H_1 < 0.14$,

- the *basin length* L_b should be equal to the length of roller L_r^* of the classical jump, and

- the *submergence degree* $(h_2 + z_t)/h_2 = 1.05$ to 1.1 is an optimum.

The optimum size of *riprap* was studied by Peterka (1958).

Tailwater Waves

Stilling basins with $2.5 < \mathbf{F}_1 < 4.5$ are prone to tailwater waves. Abou-Seida (1963) studied the (Figure 5.20(a)) relative wave height h_W/h_2, and wave steepness $\sigma_W = h_W/(gt_W^2)$ with t_W as the wave period. The inverse *wave Froude number* $\mathbf{F}_{\bar{W}}^2 = (gh_W)/V_1^2$ and the dimensionless wave period $T_W = (gt_W)/V_1$ vary with the sequent depth ratio Y and $\mathbf{F}_1$. The effect of submergence $Y = h_2/h_1$ is again significant (Figure 5.20(b),(c)).

Field Experience

Berryhill (1963) studied a number of stilling basins and arrived at the following comments:

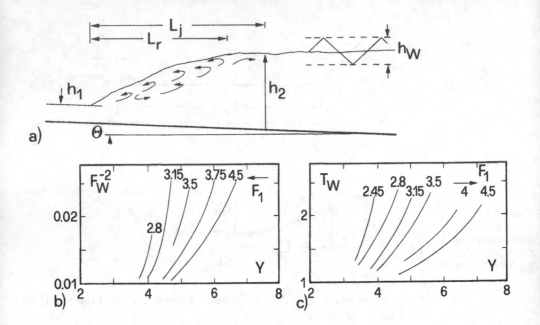

Figure 5.20 Tailwater waves in a stilling basin (a) geometry, (b) wave Froude number F_W and (c) wave period T_W in horizontal stilling basin

- the *tailwater depth* should at least be equal to the sequent depth of the classical jump,

- adequate *tailwater submergence* can shorten the basin,

- *dividing walls* can contribute to the stilling action and reduce flow concentrations,

- *cavitation damage* is increased by high approach velocity and low tailwater levels, and

- *end sills* reduce scour significantly.

Difficulties with stilling basins occur due to (ICOLD 1987):

- unacceptable scour,

- long periods of floods that damage the appurtenances,

- uplift pressure due to dynamic action,

- absence of appropriate basin lining,

- overforcing of appurtenances and cavitation damage, and

- insufficient self-cleaning of basin and resulting basin abrasion.

Cassidy, et al. (1994) arrived at the following conclusions, based on a review on damages and field experiences:

- Stilling basins should contain no chute blocks for approach velocities in excess of $30\,\mathrm{ms}^{-1}$, baffle blocks for approach velocities larger than $30\mathrm{ms}^{-1}$ should be carefully model tested.

- High, narrow baffle blocks are subjected to large fluctuating lateral forces and should not be used.

- Interference effects from adjacent blocks often prevent organization of vortex shedding from the blocks.

Stilling basins are popular and the designer's favourite choice for energy dissipation, certainly because of the knowledge and experience acquired over the years. They have proved to be a reliable hydraulic structure if the *approach conditions* and the *tailwater elevation* are within certain limits. Abrasion may become a concern for stilling basins connected to a bottom outlet.

5.3 DROP STRUCTURES AND PLUNGE POOLS

5.3.1 Basic flow features

Drop structures are used when the tailwater required for a stilling basin is not available. The approach direction is nearly vertical on to a water pool and *impact forces* are significant. Scour is thus an important aspect whereas the length of the basin is rather short.

The basic *flow types* occurring in a prismatic drop structure depend on the approach depth h_o and the tailwater depth h_u measured from the drop elevation and are (Figure 5.21):

- free-falling jet and supercritical tailwater,

- hydraulic jump if tailwater depth is smaller than drop height w,

- plunging jet flow for flow depth ratio $h_o/h_u \geq 1.17$, and

- undulating surface jet flow for $1 < h_o/h_u < 1.17$.

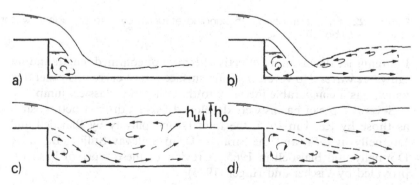

Figure 5.21
Drop structure in prismatic rectangular channel, flow types

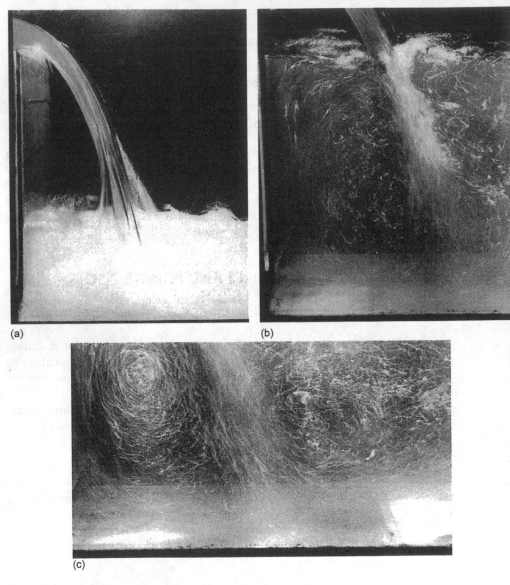

(a)   (b)

(c)

Figure 5.22 Impact structure with plunging jet for (a) low and (b) high tailwater level, (c) vorticity of flow close to bottom

Plunging jet flow is more effective because of confined length and no tailwater waves. However, a strong surface return current is generated which has a comparable intensity to the roller of a classical jump.

Several types of basins were developed using a drop structure, such as those by Rand in 1955, the Inlet Drop Spillway by Blaisdell and Donnelly in 1954, and the Straight Drop Spillway Stilling Basin by Donnelly and Blaisdell in 1965. A review of these types of basins is provided by Vischer and Hager (1995).

5.3.2 Impact structures

An impact structure receives a nearly plane and vertical jet and the energy is dissipated by jet diffusion and jet deflection. Usually, the structures have a large water cushion of thickness t_L, and the jet has a thickness t_j and an impact velocity V_1 (Figure 4.5).

From a review of experiments for the vertical jet, Vischer and Hager (1995) found for the maximum pressure p_M, the transverse pressure distribution $\bar{p}(x)$ and the rms-pressure fluctuation p'

$$p_M/(\rho V_1^2/2) = 7.4(t_j/t_L), \qquad (5.22)$$

$$\bar{p}(x)/p_M = \exp[-0.023(x/t_L)^2], \qquad (5.23)$$

$$\overline{(p'^2)}^{1/2}/(\rho V_1^2/2) = \alpha \qquad (5.24)$$

with the extremes $\alpha = +0.28$ and $\alpha = -0.04$. Results for the round turbulent jet were also presented by Vischer and Hager (1995).

5.3.3 Scour characteristics

The following refers to free jets discharging in *plunge pools*. The study of Bormann and Julien (1991) includes both drops and free jets (Figure 5.23). With h_u as tailwater depth, z_e scour depth, z_j drop height, q unit discharge, V_o approach velocity, d_n representative grain diameter, and α_j impact jet angle, the experiments yield

$$\frac{z_e + z_j}{h_o} = 0.61 \left[\frac{V_o^2}{gh_o}\right]^{0.8} \left[\frac{h_o}{d_n}\right]^{0.4} \frac{\sin\alpha_j}{[\sin(25^0 + \alpha_j)]^{0.8}}. \qquad (5.25)$$

The impact angle α_j (in rad) is for *submerged jets* (Figure 5.23(a))

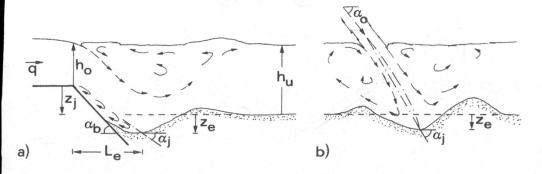

a)   b)

Figure 5.23 Scour in plunge pools for (a) submerged and (b) free jet flow

$$\alpha_j = 0.32\sin\alpha_b + 0.15\ln(1 + \frac{z_j}{h_o}) + 0.13\ln\left(\frac{h_u}{h_o}\right) - 0.05\ln[V_o/(gh_o)^{1/2}]$$

$$(5.26)$$

and for *free jets* (Figure 5.23(b))

$$\alpha_j = \alpha_o. \tag{5.27}$$

The location of the maximum scour depth is with d_{90} as the dominant grain diameter

$$L_e/h_o = 0.61\left[\frac{V_o^2/(gh_o)}{\sin(25^0 + \alpha_j)}\right]^{0.8}\left[\frac{h_o}{d_{90}}\right]^{0.4}. \tag{5.28}$$

5.4 TRAJECTORY BASINS

5.4.1 Description of structure

In a trajectory basin or *ski jump* the energy is mainly dissipated by jet dispersion. To become effective, the approach energy head should at least be 50 m, and the discharge in $[m^3s^{-1}]$ should be smaller than $250(H - H_o)$ with $H_o = 8m$ (Mason, 1982, 1993).

The trajectory basin is composed of five reaches (Figure 5.24):

① approach chute,

② deflection and takeoff,

③ dispersion of water jet in air,

④ impact and scour of jet, and

⑤ tailwater zone.

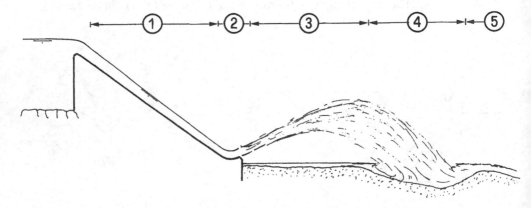

Figure 5.24 Portions of trajectory basin

The basic trajectory basin is composed of a rectangular chute of constant width and a circular-shaped take-off bucket. Special shapes involve buckets at orifice outlets in arch dams, and *flip buckets* at bottom outlets, with a curved jet trajectory in the plan view. In the following, the standard structure is considered.

5.4.2 Take-off

The take-off elevation of a ski jump relative to the dam height is about 30 to 50%. If it is too high, the velocity is not large enough for jet dispersion, and the trajectory length is too short otherwise. The *elevation of the bucket lip* should in any case be above the maximum tailwater level in order to (1) prevent material entering the bucket that causes abrasion and (2) counter cavitation damage due to submergence fluctuations.

The bucket of a trajectory basin has to deflect the water flow into the air and guide it to the proper impact location. It should operate properly under all discharges and be designed for both the static and the dynamic pressure loads. The shape of the bucket is normally a circular arc of bucket radius R_b and take-off angle α_j. The approach slope should be smaller than 4:1 and the take-off angle between 20 and 40° (Figure 5.25).

Assuming a flow with concentric streamlines of approach depth t_b and with $\mathbf{F}_o = q/(gt_b^3)^{1/2}$ as the approach Froude number, the *maximum pressure head* due to flow deflection is

$$\frac{p_M}{\rho g t_b} = \frac{t_b}{R_b} \mathbf{F}_o^2. \tag{5.29}$$

In a preliminary design, this maximum pressure head is assumed to apply along the bucket length, independent of the turning angle.

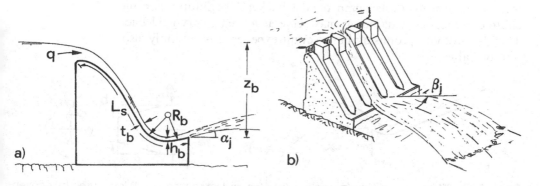

Figure 5.25 Trajectory basin (a) section, (b) transverse spread of jet

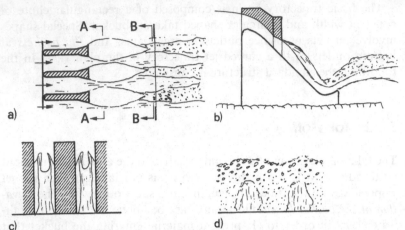

a)

b)

c)

d)

Figure 5.26
Flaring gate pier (a)
plan, (b) side view,
(c) section A-A, (d)
section B-B (Gong,
et al., 1987)

Vischer and Hager (1995) reviewed various approaches for the *bucket radius* R_b. With $H_o = V_o^2/2g$ as the approach energy head and p_M the maximum pressure, Damle, and USBR proposed, respectively,

$$R_b/t_b = (H_o/t_b)^{1/2}, \qquad (5.30)$$

$$R_b/t_b = V_o^2/(2p_M/\rho). \qquad (5.31)$$

The *flaring gate pier* is a recent development of the trajectory bucket to enhance the dispersion of flow (Figure 5.26). Those piers are located upstream from the bucket and induce air pockets in the pier wake. An air-water mixture is thus deflected into the air, instead of a compact water flow. No practical experience of this Chinese design is actually available.

The *slit-type flip bucket* was introduced by Zhenlin, et al. (1988) mainly for bottom outlets. The bucket contracts the flow, promotes air entrainment and thus the dispersion of flow (Figure 5.27). The contraction ratio must be carefully chosen to inhibit flow choking and the formation of a hydraulic jump on the bucket. The information on chute contractions (Chapter 4) may serve as a preliminary guideline. This design looks promising but relevant experience is currently also not available.

a)

b)

Figure 5.27 Slit-type flip bucket (a) section, (b) plan and jet section. (– – –) jet caused by conventional bucket (Zhenlin, et al., 1988)

5.4.3   Jet disintegration

A vast number of papers are available on the disintegration of liquid jets in air. Most of the studies involve means to improve the compactness of the jet, such as in fire nozzles or irrigation sprinklers. In contrast, the formation of a highly disintegrated jet is desirable for energy dissipation. In the extreme, the more or less compact approach flow is deflected into the air by the trajectory bucket and a highly concentrated spray falls back on to the tailwater, such that scour would be no concern. Such a high degree of disintegration is not feasible, however. The spray flow by trajectory basins was studied by Zai-Chao (1987). Figure 5.28 shows various zones to be distinguished.

The disintegration of a water jet in air can be enhanced by the:

- approach turbulence,

- approach swirl,

- approach geometry,

- counter-current wind, and

- fluid properties.

Because the jets studied under laboratory conditions are normally small in diameter, the effects of *surface tension* and *viscosity* are significant. The number of parameters that influence a liquid jet in air is so large that few general results are actually available. From the study of published information, the disintegration process can be enhanced by (Vischer and Hager, 1995):

- a non-circular cross-section to counter the compactness of the jet,

- 'roughening' the jet to increase its turbulence level (beware of damages by cavitation),

- abrupt transition from the bucket to the air, and

- adding air to the jet and creating an air-water mixture at the take-off zone.

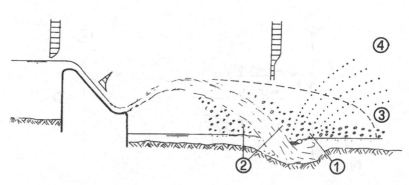

Figure 5.28
Spray flow induced by a ski-jump with ① splash drop, ② rainstorm, ③ atomization by rain, ④ atomization by wind (Zai-Chao, 1987)

All these effects should be in a relation between adding to the jet dispersion, cavitation damage as well as economy. Actually, few general guidelines are available, mainly because of lack of prototype observations. Also, the dispersion of jets as they occur at trajectory basins are governed by highly complex phenomena such as the interaction of viscosity, surface tension, air entrainment, turbulence and gravity.

Air entrainment in water jets was studied by Ervine and Falvey (1987). The lateral jet spreading was related to the turbulence number $\mathbf{T} = u'/V_o$ with u' as the rms-value of the instantaneous axial velocity and V_o as approach velocity (Figure 5.29). For a typical turbulence level of 5 to 8%, the spread of the jet was $\alpha_d = 3$ to 4%, and the inner core had a decay angle $\alpha_c = 0.5$ to 1%. A turbulent jet thus begins to break up when the inner core has completely disappeared, i.e., when the relative *breakup distance* is $L_b/D_o = 50$ to 100.

The *jet trajectory geometry* may be approximated by a conventional parabola for the local centre of gravity, on which the lateral jet spreading is added. With reference to Figure 5.30 in the $(x; z)$ coordinate system, and where α_j is the take-off angle, the trajectory follows the equation

$$z = \tan\alpha_j x - \frac{g}{2V_j^2\cos^2\alpha_j}x^2. \tag{5.32}$$

If $H_j = V_j^2/(2g)$ is the take-off velocity head with V_j as take-off velocity, the location x_M and the maximum jet elevation z_M are

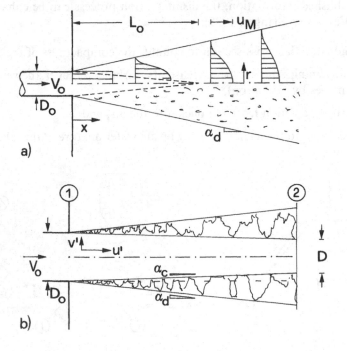

Figure 5.29
Disintegration of turbulent water jet in air (a) flow geometry, (b) spreading of jet

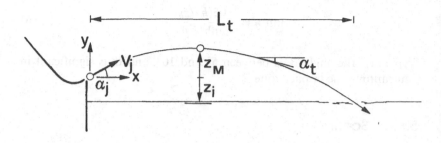

Figure 5.30
Definition of one-
dimensional jet
trajectory

$$x_M/H_j = 2\sin\alpha_j\cos\alpha_j; \quad z_M/H_j = \sin^2\alpha_j. \tag{5.33}$$

Further, the local trajectory angle α_t is

$$\tan\alpha_t = \tan\alpha_j - (x/H_j)/(2\cos^2\alpha_j) \tag{5.34}$$

and the *trajectory length* L_t is with the impact height z_i (Figure 5.30)

$$L_t/H_j = 2\sin\alpha_j\cos\alpha_j\left[1 + \left(1 + \frac{z_i}{H_j\sin^2\alpha_j}\right)^{1/2}\right]. \tag{5.35}$$

The *bucket take-off angle* α_j does not in general coincide with the bucket terminal angle α_b because of the particular pressure distribution. Orlov (1974) studied this problem and presented a diagram for the determination of $\alpha_j = \alpha_b - \alpha_o$ as a function of the deflection angle β_b and the relative bucket radius R_b/t_b (Figure 5.31). Orlov used the conformal mapping method and thus limited his results to approach Froude numbers $\mathbf{F}_o > 5(\sin\alpha_o)^{1/2}$.

The *transverse jet expansion* β_j (Figure 5.25(b)) was reanalysed by Vischer and Hager (1995) based on a Russian study. It depends mainly on the bucket flow depth relative to the fall height z_b, and the relative discharge $\bar{q} = q/(gL_s^3)^{1/2}$ with L_s as spillway length. An estimation for β_j is

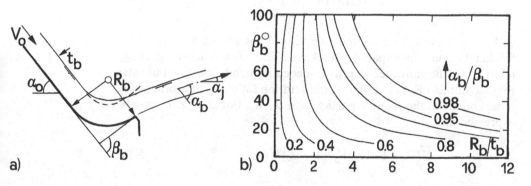

Figure 5.31 Takeoff angle $\alpha_j = \alpha_b - \alpha_o$ (a) flow geometry, (b) bucket angle ratio α_b/β_b as a function of relative bucket radius R_b/t_b for various deflection angles β_b

$$\tan\beta_j = \frac{1.05(h_b/z_b)^{1/2}}{\tanh(6\bar{q}^{1/3})}. \tag{5.36}$$

Typically, the angle β_j is between $5°$ and $10°$, and thus significant in determining the impact zone.

5.4.4 Scour

The performance of a trajectory basin is mainly related to the quality of the impact zone. If the impact area behaves differently than assumed during the design, the entire dam structure can be damaged with significant consequences to the dam safety. In the past, damage to basins of dams such as those of the Kariba (Zambia) and the Tarbela (Pakistan) power plants have led to serious concerns and considerable works after the completion of the dams only to guarantee the dam safety. Often, the geological predictions were too optimistic and the actual scour holes were much larger than predicted.

The *progress of scour* should be modelled physically for each trajectory basin. The scour material is reproduced according to the Froude similarity law and either a filler is used to simulate the cracks, or an uncohesive material is taken and only the ultimate scour area investigated. The erosion process is made of two stages: (1) the *disintegration phase* where the matrix is fractured due to the dynamic pressure action, and (2) the *transportation phase* with the rock fragments lifted entrained in the flow and deposited at the rims of the scour area.

The *scour depth* was determined both in model and in prototype situations. Figure 5.32 relates to either an overfall spillway or a trajectory spillway. Mason (1989) observed a significant effect of approach jet aeration for the latter scour depth. The aeration ratio $\beta_a = q_a/q$ between air and water unit discharges is

$$\beta_a = 0.13\left[1 - \frac{V_e}{V_i}\right]\left[\frac{H_o}{t_i}\right]^{0.45} \tag{5.37}$$

with $V_e \cong 1.1$ m/s as entrainment velocity, V_i as impact velocity, H_o as fall height and t_i as impact jet thickness. The *ultimate scour depth* $z_{e\infty}$ relative to the tailwater depth h_u depends mainly on the tailwater Froude number $\mathbf{F}_u = q/(gh_u^3)^{1/2}$, the aeration ratio β_a and is practically independent of the relative particle size d_m/h_u. For $\beta_a < 2$, it may be expressed as

$$\frac{z_{e\infty}}{h_u} = 1 + 3.4\mathbf{F}_u^{0.6}(1 + \beta_a)^{0.3}(h_u/d_m)^{0.06}. \tag{5.38}$$

The effect of tailwater Froude number is thus highly significant, and the ultimate scour depth can be reduced by adequate submergence.

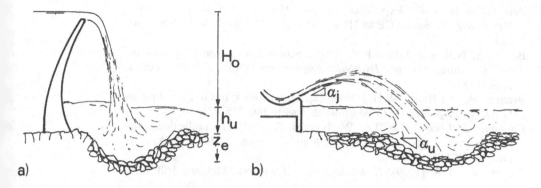

Figure 5.32 Scour in (a) overfall and (b) trajectory spillways

Figure 5.33
Photograph of
trajectory basin with
scour hole,
Nacimiento dam
after 1969 flood

The particle size has a comparatively small influence. A typical mean size of particles is $d_m = 0.25\,\mathrm{m}$. Additional information is provided by Mason (1993).

REFERENCES

Abou-Seida, M.M. (1963). Wave action below spillways. *Journal Hydraulics Division* ASCE **89**(HY3): 133–152.

Basco, D.R. (1971). Optimized geometry for baffle blocks in hydraulic jumps. *14 IAHR Congress* Paris **2**(B18): 1–8.

Beichley, G.L. and Peterka, A.J. (1959). The hydraulic design of slotted spill-way buckets. *Journal Hydraulics Division* ASCE **85**(HY10): 1–36.

Berryhill, R.H. (1963). Experience with prototype energy dissipators. *Journal Hydraulics Division* ASCE **89**(HY3): 181–201; **90**(HY1): 293–298; **90**(HY4): 235.

Bormann, N.E. and Julien, P.Y. (1991). Scour downstream of grade-control structures. *Journal Hydraulic Engineering* **117**(5): 579–594; **118**(7): 1066–1073.

Bremen, R. and Hager, W.H. (1994). Expanding stilling basin. Proc. Institution Civil Engineers *Water Maritime & Energy* **106**: 215–228.

Cassidy, J.J., Locher, F.A., Lee, W., Nakato, T. (1994). Hydraulic design for replacement of floor blocks for Pit 6 stilling basin. *18 ICOLD Congress* Durban **Q71**(R40) 599–621.

Elevatorsky, E.A. (1959). *Hydraulic energy dissipators.* McGraw Hill, New York.

Ervine, D.A. and Falvey, H.T. (1987). Behaviour of turbulent water jets in the atmosphere and in plunging pools. *Proc. Institution Civil Engineers* **83**: 295–314; **85**: 359–363.

Gong, Z., Liu, S., Xie, S., Lin, B. (1987). Flaring gate piers. *Design of hydraulic structures:* 139–146, A.R. Kia and M.L. Albertson eds. Colorado State, Fort Collins.

Hager, W.H. (1992). *Energy dissipators and hydraulic jump.* Kluwer Academic Publishers, Dordrecht.

ICOLD (1987). Spillways for dams. *ICOLD Bulletin* **58**. International Commission for Large Dams, Paris.

Mason, P.J. (1982). The choice of hydraulic energy dissipator for dam outlet works based on a survey of prototype usage. *Proc. Institution Civil Engineers* **72**(1): 209–219; **74**(1): 123–126.

Mason, P.J. (1989). Effects of air entrainment on plunge pool scour. *Journal Hydraulic Engineering* **115**(3): 385–399; **117**(2): 256–265.

Mason, P.J. (1993). Practical guidelines for the design of flip buckets and plunge pools. *Water Power and Dam Construction* **45**(9/10): 40–45.

McCorquodale, J.A. (1986). Hydraulic jump and internal flows. *Encyclopedia of fluid mechanics* **2**: 122–173. N.P. Cheremisinoff, ed. Gulf Publishing, Houston.

McCorquodale, J.A. and Khalifa, A. (1983). Internal flow in hydraulic jumps. *Journal Hydraulic Engineering* **109**(5): 684–701; **110**(9): 1508–1509.

Novak, P. (1955). Study of stilling basins with special regard to their end sill. *6 IAHR Congress* The Hague **C**(15): 1–14.

Orlov, V. (1974). Die Bestimmung des Strahlsteigwinkels beim Abfluss über einen Sprungschanzenüberfall. *Wasserwirtschaft–Wassertechnik* **24**(9): 320–321 (in German).

Peterka, A.J. (1958). Hydraulic design of stilling basins and energy dissipators. *Engineering Monograph* **25**. US Bureau of Reclamation, Denver.

Rajaratnam, N. (1965). The hydraulic jump as a wall jet. *Journal Hydraulics Division* ASCE **91**(HY5): 107–132.

Rajaratnam, N. (1967). Hydraulic jumps. *Advances in Hydroscience* **4**: 197–280. V.T. Chow, ed. Academic Press, New York.

Reinauer, R. and Hager, W.H. (1995). Non-breaking undular hydraulic jump. *Journal Hydraulic Research* **33**(5): 683–698.

Rouse, H. Siao, T.T., Nagaratnam, S. (1959). Turbulence characteristics of the hydraulic jump. *Trans. ASCE* **124**: 926–966.

Schröder, R. (1963). Die turbulente Strömung im freien Wechselsprung. *Mitteilung* **59**, H. Press, ed. Institut für Wasserbau und Wasserwirtschaft, TU Berlin, Berlin (in German).

Vischer, D.L. and Hager, W.H. (1995). *Energy dissipators*. IAHR Hydraulic Structures Design Manual **9**. A.A. Balkema, Rotterdam.

Zai-Chao, L. (1987). Influence area of atomization–motion downstream of dam. *22 IAHR Congress* Lausanne **B**: 203–208.

Zhenlin, D., Lizhong, N., Longde, M. (1988). Some hydraulic problems of slit-type buckets. Int. Symposium *Hydraulics for Large Dams* Beijing: 287–294.

References

    Wibble, R.J. and Doe, R.F. (1989) Introduction to hydrogeochemistry investigations. Wiley Inc., New York, pp. 26-30.
    White, W.B. (1988) Ground water. W.C. Brown Inc., Iowa, pp. 40-48. (Fig. 66)
    Zhang, Y.T. (1990) Isotopic profiles. A. Hoffman, Boston.
    Zhu, C. and Anderson, G.M. (1991) Environmental applications of geochemical modelling. Cambridge University Press.

Bottom outlet of Manicouga dam, Canada (*Forces* 1969)

6

Bottom Outlets

6.1 DESIGN PRINCIPLES

According to section 3.1 a bottom outlet serves various purposes such as:

- filling of the reservoir,
- drawdown of the reservoir,
- flushing of sediments, and
- flood and residual discharge diversion.

Because the velocity V at the bottom outlet is large, that is nearly as large as given by the Torricelli formula $V = (2gH_o)^{1/2}$, where H_o is the head on the outlet and g the gravitational acceleration, cavitation, abrasion and aerated flow are particular hydraulic problems. Additional *concerns* with the bottom outlet are:

- sediment flow due to reservoir sedimentation,
- gate blocked due to floating debris or sediment deposits,
- gate vibration due to high velocity flow, and
- sealing of tunnel flow due to limited air access.

Gate vibration is a particular problem in bottom outlets. It is treated in excellent works by Naudascher (1991), and Naudascher and Rockwell (1994). Problems with sediment are treated in Chapter 8. Sealing of the tailwater outlet is considered in connection with air-water flow.

A bottom outlet has to be designed such that it may be operated under all conditions for which it was planned. Usually two outlet gates are provided, namely (1) the *safety gate* or guard gate either open or closed, and (2) the *service gate* or regulating gate with variable opening.

According to Blind (1985), a bottom outlet should be provided in every dam of a certain size, particularly for emergency repair. A useful design is the combination of diversion tunnel and bottom outlet (Figure 6.1(a)). For smaller dams or for arch dams, a culvert type bottom outlet may also be considered because of the simple design (Figure 6.1(b)).

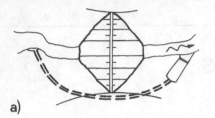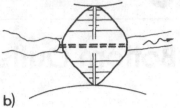

a)   b)

Figure 6.1
Arrangement of
bottom outlet.
(a) combination
with diversion
tunnel, (b) culvert
type bottom outlet

Blind (1985) presented four possible *arrangements* for the bottom outlet (Figure 6.2):

1. Diversion tunnel used as bottom outlet, with access through a shaft.

2. Bottom outlet culvert, not accessible except for minimum reservoir level, with two gates close to the inlet to shorten pressurized outlet portion.

3. Diversion tunnel used as combined spillway and bottom outlet for morning glory spillways. In the upstream sealed tunnel portion, a pipe is provided to form the bottom outlet.

4. Gravity dam with bottom outlet that is much shorter than for an earth dam.

The technical requirements for a bottom outlet may be summarized as follows (Giesecke, 1982):

• smooth flow for completely opened outlet structure,

• excellent performance for all flows under partial opening,

• effective energy dissipation at terminal outlets,

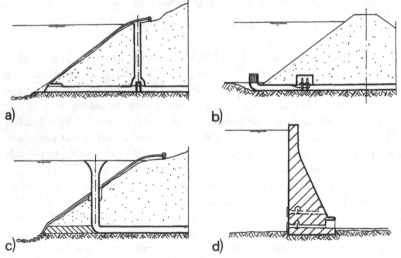

a)   b)

c)   d)

Figure 6.2
Basic arrangements
of bottom outlet

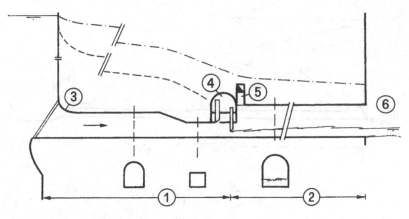

Figure 6.3
Hydraulic
configuration of
bottom outlet with
(---) pressure head
line, (-·-) energy
head line.
① Pressurized and
② free surface flow
portion, ③ tunnel
inlet, ④ gate
chamber, ⑤ air
supply, ⑥ tunnel
outlet

- structure without leakage,
- simple and immediate application,
- easy access for maintenance and service,
- economic and useful design, and
- long life.

A bottom outlet is not a structure for permanent use due to limitations regarding cavitation, hydrodynamic forces, abrasion and vibrations. It should, however, allow a complete emptying of the reservoir, as previously mentioned.

Figure 6.3 shows the *hydraulic configuration* of a bottom outlet. Note the pressurized flow upstream, and the free-surface flow downstream from the gate. At the tunnel inlet that can be provided with a trash rack, the water is accelerated to the tunnel velocity. The tunnel often has a horseshoe profile. Shortly upstream from the gate chamber, the section contracts to a rectangular cross-section to cause sufficient backpressure and to accommodate the gates. Downstream from the gate chamber, the tunnel is expanded both laterally and at the tunnel ceiling. For short tunnels, no additional aeration is needed. For long tunnels relative to the tunnel diameter, an aeration conduit discharging behind the gate chamber provides sufficient air for free-surface flow under practically atmospheric pressure. The air supply conduit has to be designed so that the gate chamber is safe against submergence from the tunnel.

It is imperative that *submergence* of the bottom outlet is inhibited. Accordingly, the transition from pressurized to free surface flow has to be located exactly behind the gate chamber. This condition requires sufficient aeration, and a tunnel large enough that surging flow may not develop in the downstream tunnel. Also, the discharge is then fully controlled with the gates.

The hydraulics of a bottom outlet across an earth dam is sketched in Figure 6.4. In Figure 6.4(a), the tunnel has a constant diameter, and no

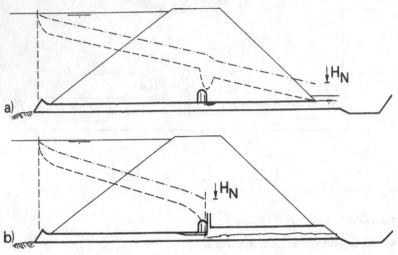

Figure 6.4
Bottom outlet
across earth dam
(a) fully pressurized
flow, (b) free-
surface flow
downstream from
gate chamber.

aeration is provided. Accordingly, *pressurized flow* results for a large discharge. Therefore, the discharge depends mainly on the tunnel section F, and not on the gate section f. The latter adds only to the loss of head, i.e. it involves an additional energy loss. In contrast, for a *free-surface flow* downstream from the gate chamber, the gate section f controls the discharge linearly.

In practice, this feature of bottom outlets has sometimes been overlooked, particularly with smaller structures. This can cause costly adaptions when such dams have to be rehabilitated due to safety reasons.

6.2 GATE TYPES AND DESIGN REQUIREMENTS

Figure 6.5 shows the most frequent types of outlet structures used (Task Committee, 1973; Giesecke, 1982; Sagar, 1995). (a) *Wedge-gate* moved vertically, with complete gate sealing only at fully closed position. Instead of a wedge, two interconnected and displacing plates are also in use. A disadvantage of this gate type is the gate slots as for roller gates where complex hydraulic currents may set up and sediment may enter. Flush conduits should be provided to ensure complete closure. (b) *Slide gate* as the common bottom outlet gate type. Wheel gates are considered for large heads and they are known to be less prone to vibration. (c) *Radial* or *sector gate* does not use gate slots, and sealing is simple. The forces are concentrated to the gate trunnions, and the abutments are highly stressed. Actually this is one of the favourite arrangements for bottom outlets, and of all applications of gate flow with large discharges in general. (d) *Hollow jet valve* comparable to the ring valve, but with an aeration device to break the compactness of outflow jet. (e) *Ring valve* as an element that can be

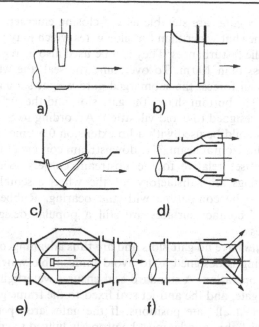

Figure 6.5
Types of
valves and
gates for
outlet
structures

displaced axially and has again an excellent hydraulic performance. The valve may be regarded as an extension of the pipe which can be displaced axially. Accordingly a ring jet develops with small losses due to streamlining. It can be used as either a regulating or a safety device for high pressures without problems in leakage. The ring valve is suited as terminal structure of the bottom outlet. (f) *Conic valve* as a simple and most effective device with a displacing element. The hollow jet is dispersed significantly in the atmosphere and problems with cavitation are absent due to aeration. Also, vibrations are not a concern for all degrees of opening. Maintenance is simple because all parts are located outside from the bottom outlet. Often, a stilling basin is not needed but the outlet should be covered due to spray action.

According to Sagar (1979), vibration, non-closure, cavitation and abrasion are the main reasons for failure of high head gates. In the following, some features of modern gate experiences are presented.

Slide gates should have a crest sloping under 45° to inhibit vibration and to reduce the downpull forces. Also, the offsets due to the gate slot should be minimized. Depending on the velocity and the sediment concentration, a bottom outlet should be lined. The tolerances should closely be checked for all gates. All parts should not only be carefully welded to avoid distortion, but often need stress-relieving prior to machining. Slide gates are suitable for heads up to 200 m, normally in a tandem arrangement. They can be arranged at the downstream end of the outlet or within the conduit up to sizes of 10m². The gate area should be somewhat smaller than the approach conduit to ensure positive water pressure when the gate is fully opened. These gates are not suitable in outlets where self-closing is required for quick shut-off in an emergency.

Fixed wheel gates are suitable as self-closing emergency gates. The slots are somewhat wider than for slide gates which may give concerns with hydraulic disturbances. They can be used also as regulating gates for heads less than 100 m. To overcome the seal, the wheel and the guide frictional forces, the submerged gate weight should be at least 30% larger. The bottom shape, the gate slot, and the air vents should carefully be designed to avoid vibration. According to Sagar (1979), a vertical lip should be used with a lip extension 0.5 times the depth of the bottom horizontal beam. The downstream corners of the gate slots should be offset relative to the upstream corners. Self-lubricating bronze bushings are satisfactory for the wheel assemblies, and the grease has to be compatible with the bearing. Rubber seals with teflon coated contact surfaces are still a popular design to ensure watertightness.

Radial gates have no gate slots and are thus advantageous, provided the top sealing is designed as to avoid undesirable water jets during partial gate operation. A standard seal ensures watertightness for the fully closed gate, and the anti-jet seal fixed to the frame gives a water-tight contact at all gate positions. If the gates are operated with a hydraulic hoist, the cylinder must be properly hinged to ensure smooth gate operation without undue stresses in the stems and lifting mechanism. The gate trunnion has to be protected against corrosion and debris.

Based on failures of high-head gates, Sagar (1979) recommends:

- If bonnet covers are used, they should be bolted down to bonnets or gate frames to distribute the water load on the concrete surroundings, and direct load transfer is avoided.

- Adequate *aeration* should be ensured downstream of the gate to inhibit cavitation and vibration. The inlets for the air vents are located such as to provide an aeration uninterrupted by tailwater, waves or debris.

- Open gate shafts should never be submerged due to significant vertical flow.

- The top and side seals should be effective for all positions of a regulating gate.

As an example of gate flow optimization, consider Figure 6.6 with (a) the original and (b) the final design. It refers to a project for the drinking water supply of Athens, Greece. The head on the wheel gate is 125 m, the gate area 3.1m × 3.1m and the maximum discharge $125 m^3 s^{-1}$. Hydraulic modelling was used to answer questions relating to the:

- shape of gate geometry,
- hydrodynamic forces on the gate,

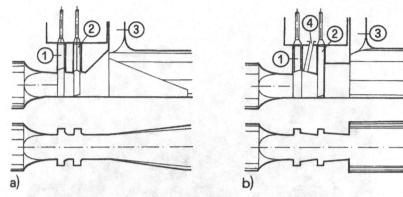

Figure 6.6
Bottom outlet in section and plan view (a) original and (b) final design (Dammel, 1977).
① safety gate,
② service gate,
③ tailwater aeration conduit,
④ intermediate aeration.

- flow conditions in the gate slots, and

- requirements of aeration.

The inlet geometry, as defined by the US Army Corps of Engineers (1957) was found satisfactory for transition from the 6 m pipe to the 3.1 m square section. The transition between the gate slots and the tailwater tunnel of horseshoe-shape induced low pressures, and linear transitions were found to perform better than ellipses. Also, the cross-section abruptly expanded to 4.5 m behind the service gate instead of a linear height increase. Accordingly, the jet would spring clear from the gate crest, and unique wall pressure conditions developed.

If the regulating gate should be inhibited for any reason, the safety gate must close without problems. Then, vibrations are a particular concern due to an interrelation between the two gates (Naudascher and Rockwell, 1994). An appropriate aeration between the tandem gates was found essential.

Glen Canyon dam, the largest and most important feature of the Colorado river project, has a height of 216 m (Figure 6.7). The diversion tunnel of 12.5 m diameter was plugged after use and serves now as the bottom outlet with three parallel conduits extending through the plug. The flow is regulated by tandem slide gates 2.1m × 3.2m, and the bottom and sides of the downstream conduits are lined with 19 mm steel. Air to each regulating gate is supplied through the space above the free water surface in the downstream conduits.

After two years of operation, the tunnels were inspected and considerable erosion and cavitation damage was found due to depressions in the invert profile and along the steel lining (Wagner, 1967; Falvey, 1990). Other damage was also found in the top between the service and the safety gates. The major damage to the gates and the conduit liner was attributed to *irregularities or misalignment* of the fluidway surfaces. Irregularities must be controlled by rigid manufacturing and installation tolerance. It was concluded that the prototype paints must be carefully applied to avoid surface roughness that may enhance

a)

Figure 6.7
Glen Canyon dam
(USA) (a) damages
of tunnel (Falvey,
1990), invert.

cavitation. It should also be considered that the roughness may be initiated by *abrasion* due to sediment flowing in the bottom outlet. The abrasive action may be increased at locations of a hydraulic jump where the material circulates like in a *ball-mill*. Therefore, all foreign material should be removed periodically and kept from entering from a hydraulic jump basin.

The *selection of a gate* should take into account past experience with successful designs and manufacturing capacities. According to Erbiste (1981) gate designers appear to be rather conservative in their approach, preferring whenever possible to adopt a type of gate already proven in practice, unless a new proposed type can be shown to have

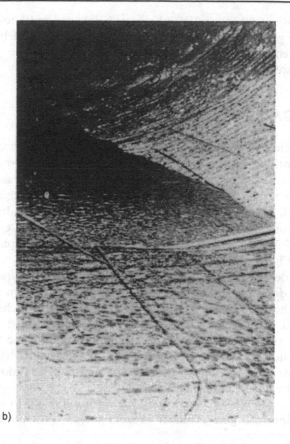

Figure 6.7
(Continued)
Glen Canyon dam,
b) view of tailwater
invert

b)

greater safety and economy. The continuous development of hydro-electric projects demands equipment of increasing size.

Based on a review of existing gates including more than 4000 references, Erbiste was able to relate various types of gates to the *gate dimensions*. Figure 6.8(a) refers to high pressure gates and indicates that for heads larger than 150 m, slide gates with a restricted area of 20m² are used exclusively. Fixed wheel and tainter gates are applied approximately to the same extent and caterpillar gates are more suitable than fixed wheel gates for heads larger than 100 m.

Figure 6.8
a) Area (m²) of high-head pressure gates as a function of head (m) for ① slide,
② caterpillar, and ③ tainter/fixed-wheel gates. (b) height of underflow gates in (m) as a function of width (m) for ① slide,
② fixed-wheel, and ③ tainter gates (Erbiste, 1981)

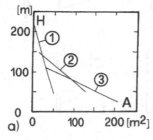

a)

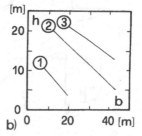

b)

In Figure 6.8(b) the relation between height and width for underflow spillway gates is shown. Currently, the tainter gate is widely used. Due to the reduction of the friction forces the width of the gates could progressively be increased in the past.

6.3   GATE VIBRATION

Naudascher and Rockwell (1994) have distinguished between the predominantly extraneously induced exiting and instability-induced exiting. Examples of extraneously induced vibration are the skinplate of a sluice gate, the tandem gate during emergency closure, and the tainter gate with two-phase flow. Predominantly instability-induced exiting occurs for high-head leaf gates with an unstable jet, a cylinder gate with a bistable underflow, or leaf gates involving impinging shear layers (Figure 6.9).

In order to protect a gate against flow-induced vibration in the *vertical* direction, either sufficient damping or a bottom shape from which the flow remains unseparated or stably reattached must be provided. Naudascher and Rockwell (1994) have pointed to the effect of gate crest geometry on the latter condition, and some typical configurations are considered. Figure 6.10 lists crest shapes where vertical vibration may or may not occur. The crest height ratio a/e should always be much larger than unity.

Other examples of vortex induced vibration relate to a gate withdrawn into the gate chamber (Figure 6.9(a)), pressure fluctuations

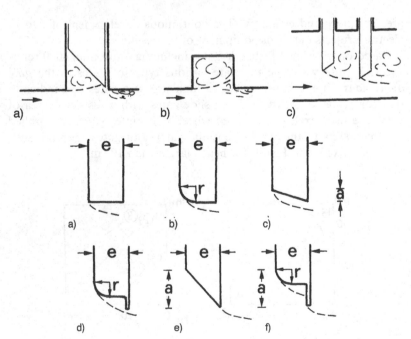

Figure 6.9
Possible sources of gate vibration with impinging shear-layer instability (Naudascher and Rockwell, 1994)

Figure 6.10
Crest shapes (a) to (d) unstable, and (e), (f) stable for vertical gate vibrations in bottom outlet.

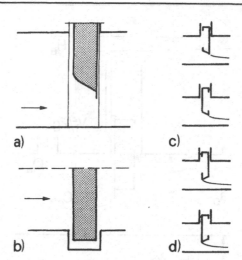

Figure 6.11
Common geometry
of high-head leaf
gate (a) side view,
(b) plan view,
(c), (d) Possible top
seal and skin plate
arrangements
(Naudascher, 1991).

within a rectangular gate slot (Figure 6.9(b)), and a tandem gate arrangement with the service gate stuck (Figure 6.9(c)).

Horizontal vibrations of gates with underflow are not as dangerous as vertical excitations. The stable lip shapes are comparable to that of Figure 6.10(e), provided the thickness of the lip is small and the gate passes quickly through small degrees of openings.

The most dangerous mechanisms for movement-induced excitation of gates and gate seals are those involving *coupling* with fluid flow pulsations (Naudascher and Rockwell, 1994). They may occur if a seal arrangement gives rise to flow *through* a leak channel with an upstream constriction. The latter should be eliminated or moved towards the downstream end of the flow passage.

The *leaf gate* is the most widely used high-head gate. From all types of gates according to 6.2 it causes the greatest problems with hydro-dynamic loading and downpull. The pressure along the crest of the leaf gate is dramatically reduced due to the high efflux velocities. The downpull thus often exceeds the weight of the gate. Figure 6.11 shows side and plan views of a common leaf gate, together with typical arrangements for sealing. The most feasible arrangement is reproduced in Figure 6.11(c) top with seal and skinplate of the downstream side. Despite the advantage of downpull reduction, gates with upstream seals can only be used for low heads due to vortex action in the gate slots.

6.4 HYDRAULICS OF HIGH-HEAD GATES

6.4.1 Discharge equation

Gate flow may be either free or submerged. For *free gate flow,* the space behind the gate is filled with air of pressure head h_a. If the efflux

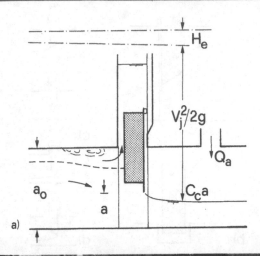

a)

b)

c)

Figure 6.12
Definition sketch for
free gate flow. (a)
Experimental setup,
(b) overall view, (c)
detail of gate flow
(VAW 45/72/12)

is into the atmosphere, $h_a = 0$. Submerged gate flow in bottom outlets should be avoided, as discussed in 6.1. Based on energy considerations, the underflow discharge Q of a gate is (Naudascher, 1991)

$$Q = C_c ab[2g(H - H_e - C_c a - h_a)]^{1/2} \qquad (6.1)$$

where C_c is the *contraction coefficient*, a the gate opening, b the gate width, and $H - H_e$ the head on the gate with H_e the energy loss from the entrance to the gate section (Figure 6.12).

The overflow discharge Q_o of a high-head gate is

$$Q_o = C_o a_3 b[2g(H - H_e - H_{ce} - h_s - h_a)]^{1/2} \qquad (6.2)$$

where C_o is a discharge coefficient for the seal system, a_3 the width between the slot and the gate, H_{ce} the head loss between the approach flow and the upstream face of gate, and h_s the vertical distance between the datum and the section of jet issue.

The *discharge coefficient* C_d of gate flow is the ratio

$$C_d = Q/(2g\Delta H)^{1/2} \qquad (6.3)$$

with ΔH as difference head according to Eq.(6.1). The parameter C_d depends on a number of parameters such as the relative gate opening $\eta = C_c ab/A_o$, with $A_o = a_o b_o$ as the approach section, the loss factor, the aspect ratio and the approach velocity distribution. Accordingly, one should rather use the contraction coefficient C_c than the lump parameter C_d.

The *contraction coefficient* C_c depends on the efflux Froude number $\mathbf{F}_j = V_j/(gC_c a)^{1/2}$ provided $\mathbf{F}_j < 4$. For large Froude numbers as are typically encountered in prototype high-head gates neither this free surface effect nor the effect of viscosity have to be accounted for, and one may consider a potential flow with the gravity effect absent, i.e. the gate geometry has the significant effect on C_c. Figure 6.13 shows contraction coefficients for a typical leaf gate geometry with a relative crest radius $R = r/e = 0.4$ and $a_o/e = 6$.

Figure 6.13
Contraction coefficient C_c (a) Definition sketch, effect of (b) gate thickness $A = a/e$ ($\theta = 30°, R = 0.4, f/e = 0$), (c) crest angle θ, (d) relative offset length $F = f/e$ and (e) relative gate radius $R = r/e$ as a function of gate opening $A = a/e$ for $r/e = 0.4$, $a_o/e = 6$ (Naudascher, 1991)

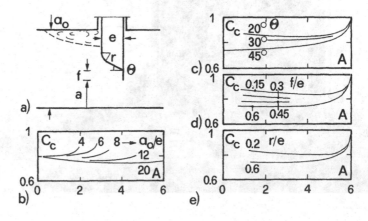

The *gate overflow* is complicated by corner eddies. An analysis is presented by Naudascher (1991) accounting for the detailed flow structure between the body of the gate and the gate chamber. The *downpull* of a gate is also a specialized domain of hydrodynamic forces and will not be treated here, because of Naudascher's exhaustive treatment. For important gates, model experiments are necessary and recommended.

6.4.2 Gate slots

The performance of slide and wheel gates depends highly on the gate slots in which they are supported. Ball (1959) has presented basic information for gate slots, and damages on both the gate and the outlet structure should not occur. The tests were made with the gate fully open and for pressurized flow, and it was demonstrated that this corresponds to the determining flow configuration for the gate slots. Among various slot geometries, the configuration with abrupt upstream and downstream corners was retained provided the downstream corner was *recessed* to obtain a gradual convergence towards the tailwater wall (Figure 6.14(a)). A final design was presented by the United States, Army Corps of Engineers (1968). It involves a rounded slot geometry 58% deep of the gate length, as shown in Figure 6.14(b)).

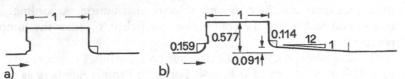

Figure 6.14
Gate slot (a) without and (b) with downstream offset (recommended design). All lengths are multiplied by the slot length

a)

Figure 6.15
Effect of guide rail in gate slot (a) with

Figure 6.15
(Continued)
(b) without rail (*Proc. Institution Civil Engineers* 1985, 79(2): 755)

b)

Figure 6.15 refers to photographs of gate slots with and without the presence of a guide rail. The system of vortices is seen to be significantly different for two otherwise identical configurations.

6.5   AIR ENTRAINMENT

Whereas the flow is pressurized upstream from the gates of a bottom outlet, free surface flow may develop in the tailwater of the gate, depending on the flow characteristics in the outlet tunnel. Because free gate outflow greatly reduces the potential of gate vibration and cavitation damage, a bottom outlet should always be designed for *free surface flow*.

The aeration of flow may originate from three different sources (Figure 6.16):

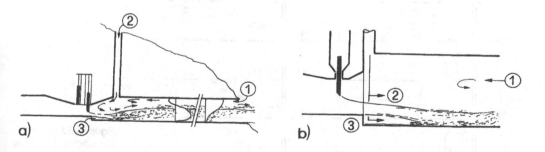

Figure 6.16   Air entrainment by gated outflow in bottom outlet tunnel (a) overall view, (b) gate chamber with ① counter-current air from tunnel, ② air supply conduit for surface aeration, ③ bottom aerator

① *Tunnel outlet* in a counter-current air flow along the outlet roof,

② *Air supply* conduit by which the underpressure of the surface air is reduced, and

③ *Bottom aerator* that counters problems with cavitation damage.

Due to limited air availability the air flow in a bottom outlet is more complex as compared to chute aerators. The hydraulics of air-water flows in bottom outlets is actually not yet fully understood, and hydraulic modelling based on sufficiently large scale models is recommended. The Froude similarity law governs such flow, and successful studies have been conducted in the past (Giezendanner and Henry, 1980; Anastasi, 1983). These and other studies stressed the effect of flow type on the air entrainment characteristics.

Sharma (1976) has introduced a classification of *flow types* for bottom outlets without bottom aerator (Figure 6.17):

1. *Spray flow* for relative gate opening below 10%, with an extremely high air entrainment,

2. *Free flow* as typical for bottom outlets, and accompanied by features of supercritical flow, such as shockwaves and two-phase flow,

3. *Foamy flow* for a tunnel almost full with an air-water flow,

4. *Hydraulic jump* with a free surface tailwater flow due to a tailwater submergence,

5. *Hydraulic jump* with transition to pressurized tailwater flow, and

6. *Fully pressurized flow* caused by a deep tailwater submergence.

In applications, only types 1. to 3. should occur because of the dangerous surging conditions otherwise.

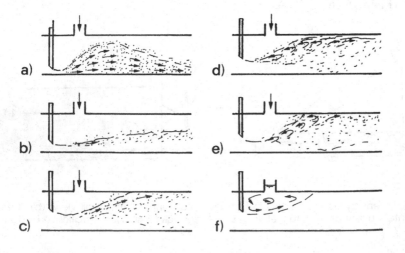

Figure 6.17
Classification of flow types in bottom outlets *without* bottom aerator (Sharma, 1976)

A preliminary design of air entrainment was presented by Rabben and Rouvé (1985). A fictitious cross-sectional air conduit section $A_a^* = A_a(1 + \Sigma\xi_i)^{-1/2}$ was introduced where $\Sigma\xi_i$ is the sum of all head losses from the atmosphere to the gate chamber. For *spray flow* (Figure 6.17(a)) with a relative gate opening below 6% and $\mathbf{F}_c \geq 20$, the *air ratio* $\beta = Q_a/Q$ obtains

$$\beta = (A_a^*/A_d)\mathbf{F}_c. \tag{6.4}$$

The tailwater tunnel section is A_d, and $\mathbf{F}_c = q/[g(C_c a)^3]^{1/2}$ is the Froude number relative to the contracted gate flow depth. The air discharge is Q_a, and the water discharge is $Q = qb_g$ with b_g as gate width, g is the gravitational acceleration, and a is the gate opening.

For *free gate flow* (Figure 6.17(b)) with a relative gate opening in excess of 12%, and $\mathbf{F}_c \leq 40$, a similar expression for the air ratio was found as

$$\beta = 0.94(A_a^*/A_d)^{0.90}\mathbf{F}_c^{0.62}. \tag{6.5}$$

For tunnel flow with a *hydraulic jump* the tailwater depth or the corresponding pressure head for pipe flow must be computed with a backwater curve, based on the outflow submergence. For conditions analogous to free flow (Figure 6.17(d)), the air ratio is approximately

$$\beta = 0.019(A_a^*/A_d)\mathbf{F}_c. \tag{6.6}$$

This is much lower than for spray flow. Alternative approaches are due to Bollrich (1963) and Sharma (1976). In all approaches, the *length of tunnel* is not accounted for, and the previous relations have been established for short tunnels. For longer tunnels, less air is entrained because of deaeration processes. Model studies based on the Froude similarity law should then be conducted.

Bottom outlets with an *aerator* are governed by complex phenomena. Figure 6.18(a) shows a definition plot. Rabben and Rouvé (1985) have presented a preliminary design procedure, but additional experiments have to be conducted before application to design. A summary of the preliminary design procedure is given by Sinniger and Hager (1989).

The effect of relative *tunnel length* and a standardization of the jet aeration have not been investigated yet. The actual tendency is to aerate the jet not only from the bottom, but also from the sides just beyond the gate section. A design procedure is not available and model observations on a sufficiently large scale model are recommended. Some information on recent prototype observations are due to Volkart and Speerli (1994) and Lier and Volkart (1994).

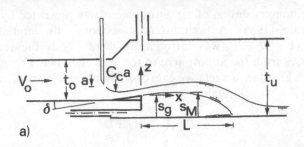

a)

b)

Figure 6.18
Bottom aerator in
bottom outlet (a)
schematic flow
structure and air
supply, (b)
photograph of slide
gate at Panix dam,
Switzerland. Gate
opening 10 mm, for
a pressure head of
50 m. Note shock
waves due to gate
slots and spray
development due
to small opening.
Flow direction from
bottom to top (VAW
43/62/28)

6.6 CAVITATION

6.6.1 Description

Cavitation is defined as the formation of a bubble or void in a liquid. In water flow the bubble is usually filled with vapour. When water is heated the temperature increases and the vapour increases also. Boiling will occur when the vapour pressure equals the local pressure. The boiling temperature is a function of the ambient pressure, a decrease of which makes water boil at lower temperatures.

Boiling is the process of passing from the liquid to the vapour state by increasing the temperature, while cavitation occurs by decreasing the local pressure under constant temperature. The local pressure reduction can be caused in fluid flow by a decrease of the total energy head because of increase in elevation, by a local increase of velocity (e.g. due to contraction) and also by turbulence, vortices or large scale separation. The water flowing in hydraulic structures contains air bubbles of various sizes and with numerous impurities. These conditions are necessary to initiate cavitation, and determine the potentials of damage and noise.

When the pressure in a fluid flow is continuously reduced due to velocity increase such as over a spillway crest, a critical point is reached where cavitation begins. This is called *incipient cavitation*. Similarly, if cavitation exists and the velocity is reduced, a critical condition is reached where cavitation starts to disappear, and this is referred to as desinent cavitation. Both incipient and desinent cavitations do not occur at the same flow conditions.

The hydrodynamic parameter describing the cavitation process is the *cavitation index*

$$\sigma = \frac{p - p_o}{\rho V_o^2/2}. \tag{6.7}$$

Herein, p is the local pressure, and p_o and V_o are the reference pressure and velocity, typically of the upstream flow. To avoid ambiguities, both vapour and reference pressures are referenced to absolute zero pressure. Although one parameter such as the cavitation index is not able to describe the various complex flow situations, it is a useful quantity to indicate the *state of cavitation*. For example, for flow past a sudden into-the-flow offset, cavitation will not occur if $\sigma > 1.8$. If $\sigma < 1.8$ more and more cavitation bubbles form that can be detected visually as a fussy white cloud. For even lower cavitation index the cloud that consists of individual bubbles suddenly forms long supercavitating vapour pockets (Figure 6.19).

Figure 6.19 Development of cavitation for flow past a sudden into-the-flow offset with $\sigma =$ (a) 3, (b) 1.8 with incipient cavitation where cavitation bubbles occasionally can be observed, (c) $0.3 < \sigma < 1.8$ with developed cavitation and (d) $\sigma < 0.3$ supercavitation (Falvey, 1990). Reference quantities referred to upstream flow

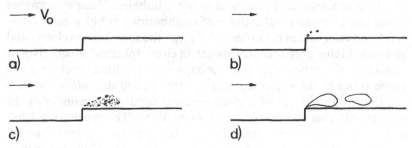

6.6.2 Bubble collapse dynamics

To simulate the bubble collapse dynamics the compressibility of water has to be considered. The bubble collapse consists of phases in which the bubble diameter decreases from the original size when the pressure is increased, reaches a *minimum size* and then grows or rebounds, as shown in Figure 6.20(a). The process is repeated for several cycles with the bubble diameter decreasing during each cycle until it finally becomes microscopic in size. During the rebound phase a shock wave forms with the shock celerity equal to the speed of sound in water. At a distance of two times the initial bubble radius from the

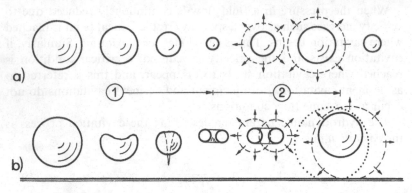

Figure 6.20
Collapse of an
individual bubble
(a) in a quiescent
fluid, (b) near a
boundary.
① collapse,
② rebound

collapse centre, the pressure intensity is about 200 times the ambient
pressure at the collapse site. The time for a bubble to collapse
depends mainly on the initial bubble radius and is of the order of
some microseconds. If the bubble collapses near a boundary, the
boundary reflects the flow towards the bubble and causes an asymmetric collapse. As shown in Figure 6.20(b) this causes one side of the
bubble to deform into a *microjet* that penetrates the opposite bubble
side. The velocity of this microjet is large and the shockwave generates
pressure so high that it may cause *cavitation damage* to the solid
boundary.

If more than one bubble is present, the collapse of the first will
produce shockwaves that radiate to other bubbles. These shockwaves
cause the unsymmetrical collapse of neighbouring bubbles. Such ultra-
jets have velocities in the order of 1.5 times the sonic water velocity and
generate higher pressure intensities than either spherical shockwaves or
microjets. If a bubble in a swarm collapses the resulting shockwave will
cause other bubbles to collapse in its vicinity, and the *collapse process*
continues in the form of a chain reaction until the remainder of the
swarm collapses simultaneously (Falvey, 1990). The synchronous collapse of a bubble swarm produces a higher pressure intensity than the
random collapse of individual bubbles in a swarm. Unfortunately, a
theory does not yet exist for the prediction of the pressure magnitude
generated by a bubble swarm.

Damage caused by a group of bubbles trapped by a *vortex* can be
many times larger than caused by the collapse of an individual bubble.
Shear flows as occur typically along a boundary thus collect bubbles
on their axes.

6.6.3 Cavitation characteristics

A flow surface irregularity may be either an isolated roughness or a
uniformly distributed roughness. Typical isolated roughnesses include
(Figure 6.21):

- offset into-the-flow,
- offset away-from-the-flow.
- grooves, and
- protruding joints.

Cavitation is formed by turbulence in the *shear zone* due to the sudden change of flow direction. Depending on the roughness shape, cavitation bubbles collapse either within the flow or near the flow boundary.

Figure 6.22(a) refers to the cavitation characteristics of the into-the-flow offset. For an approach velocity of, say 20ms^{-1}, and an offset of only 0.8 mm, the pressure head for incipient cavitation to occur is 7.7 m above vacuum. Accordingly, for pressure over the atmospheric pressure (+10m) cavitation is no problem. However, if the offset has double height, i.e. 1.6mm, then the pressure has to be at least 17.8 m above vacuum.

Figure 6.22(b) refers to a generalized plot of the *cavitation index* $\sigma_i = (h - h_v)/(V_o^2/2g)$ for incipient cavitation, with h as absolute pressure head and h_v as the vapour pressure head. Falvey (1990) has presented a comprehensive set of data for other surface irregularities, including holes, grooves, uniform roughness, and combinations of uniform and isolated roughnesses. Note that the *superposition principle* is applicable, at least to lowest order.

Figure 6.21
Isolated roughness elements with vapour cavities and damage zones, details see text.

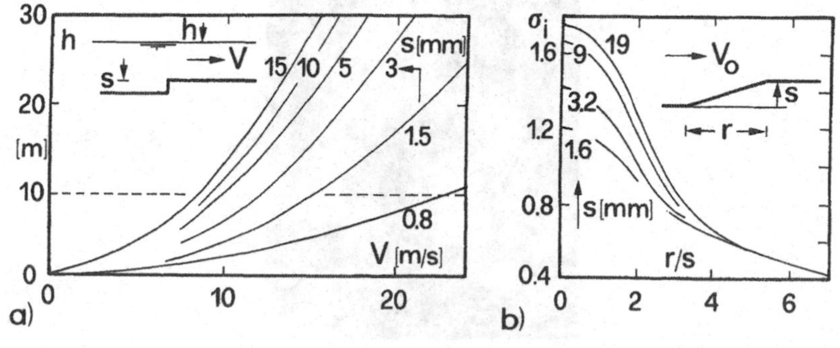

Figure 6.22
Incipient cavitation for (a) abrupt offset and (b) cavitation index σ_i for chamfered offset (Falvey, 1990).

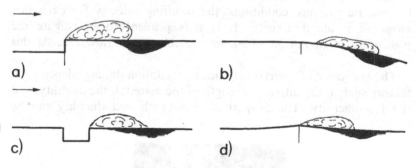

6.6.4 Cavitation damage

Cavitation as such is no danger for any structure. If cavitation occurs close to flow boundaries, then cavitation damage may happen, and the structure can seriously be damaged. The *surface damage* begins at the downstream end of the cloud of collapsing cavitation bubbles. After some time, an elongated hole is formed in the boundary surface. As time progresses, the hole gets larger with high velocity flow impinging on the downstream end of the hole. In contrast, this flow is able to create high pressures within the minute cracks around individual pieces of aggregate. Pressure differences between the impact zone and the surrounding area are able to break away parts of the aggregate and sweep it away by the flow. This process can be regarded as erosion but the loss of material due to cavitation is different.

Figure 6.23 shows a hole 11 m deep in the spillway invert of the left tunnel of Glen Canyon dam, USA. Concrete lumps were found attached to the end of the reinforcing steel. At this stage a high velocity flow acting on the lumps rip reinforcement bars from the concrete even though the steel may be imbedded as deep as 150 mm. After the tunnel lining was penetrated, erosion continued into the underlying rock. When damage penetrated the liner, the integrity of the structure became a major concern.

Based on a typical value of 2 for the cavitation index and for barometric pressure conditions, the resulting velocity for *cavitation inception* is about 10ms^{-1}. It is thus prudent to investigate the possibility of cavitation whenever the velocity of flow exceeds this limit.

The resistance of a certain surface to cavitation damage depends on factors such as the ultimate strength of the material, the ductility, and the homogeneity. The properties of strength and ductility can be

Figure 6.23
Cavitation damage
to Glen Canyon
dam tunnel spillway
(Falvey, 1990)

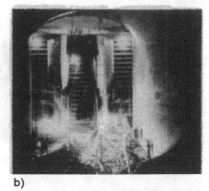

a)   b)

Figure 6.24 Damage to Tarbela tunnel 2 due to cavitation. Views from (a) upstream and (b) downstream (Kenn and Garrod, 1981)

combined into the parameter known as *resilience*, i.e. the area under the stress-strain curve of a material. A comparative cavitation resistance of various materials indicates a relatively poor resistance of concrete and polymer impregnated concrete, followed by aluminium or copper and finally by stainless steel. There exist no materials, however, that are not prone to cavitation damage. The effect of exposure time is discussed by Falvey (1990). Figure 6.24 shows typical damages on tunnel 2 of Tarbela dam, Pakistan.

6.6.5 Control of cavitation

Cavitation damage can be controlled by two methods:

- Control of cavitation index by geometry,
- Control of cavitation damage by aeration.

In a specific case, the irregularities of a spillway or a bottom outlet have to be estimated, based on accuracy specification, quality of material and site considerations. Then, the *concept of incipient cavitation* has to be applied and it must be determined whether or not cavitation damage occurs. Such computations involve a prediction of the surface profiles based on backwater curves, the determination of the pressure head curves for all relevant discharges, and the computation of the index curves. If there are locations where cavitation is predicted, then the geometry of the spillway or the smoothness of the boundary have to be improved. If both approaches fail, the flow has to be aerated, as described previously. Note that the capacity of a spillway with an aerated flow has to be larger because of the increase of air-water discharge. Details of spillway aerators are given in Chapter 4.

Figure 6.25 Schiffenen dam, Fribourg, Switzerland. Combined surface and bottom outlets, total discharge capacity 1000m³s⁻¹ (Courtesy of Gicot Geotechnique, Fribourg)

REFERENCES

Anastasi, G. (1983). Besondere Aspekte der Gestaltung von Grundablässen in Stollen. *Wasserwirtschaft* **73**(12): 501–509 (in German).

Ball, J.W. (1959). Hydraulic characteristics of gate slots. *Journal Hydraulics Division* ASCE **85**(HY10): 81–114; **86**(HY4): 121–126; **86**(HY5): 133–143; **86**(HY6): 87–89; **87**(HY1): 155–163.

Blind, H. (1985). Design criteria for reservoir bottom outlets. *Water Power & Dam Construction* **37**(7): 30–33.

Bollrich, G. (1963). Belüftung von Grundablassverschlüssen. *Wissenschaftliche Zeitschrift* Dresden **12**(6): 1709–1717 (in German).

Dammel, W. (1977). Einige Aspekte zur Konstruktion von Tiefschützen. *Bauingenieur* **52**: 353–355 (in German).

Erbiste, P.C. (1981). Hydraulic gates: the state-of-the-art. *Water Power & Dam Construction* **33**(4): 43–48.

Falvey, H.T. (1990). Cavitation in chutes and spillways. *Engineering Monograph* **42**. US Bureau of Reclamation, Denver.

Giesecke, J. (1982). Verschlüsse in Grundablässen – Funktion und Ausführung. *Wasserwirtschaft* **72**(3): 97–104 (in German).

Giezendanner, W. and Henry, P. (1980). Vidange de fond à grande vitesse. *Ingénieurs et Architectes Suisses* **106**: 387–394 (in French).

Kenn, M.J. and Garrod, A.D. (1981). Cavitation damage and the Tarbela tunnel collapse of 1974. *Proc. Institution Civil Engineers* **70**(1): 65–89; **70**(1): 779–810.

Lier, P. and Volkart, P.U. (1994). Prototype investigation on aeration and operation of the Curnera high head bottom outlet. *18 ICOLD Congress* Durban **Q71**(R36): 535–553.

Naudascher, E. (1991). Hydrodynamic forces. *IAHR Hydraulic Structures Design Manual* 3. Balkema, Rotterdam.

Naudascher, E. and Rockwell, D. (1994). Flow induced vibrations. *IAHR Hydraulic Structures Design Manual* 7. Balkema, Rotterdam.

Rabben, S.L. and Rouvé, G. (1985). Belüftung von Grundablässen. *Wasserwirtschaft* **75**(9): 393–399 (in German).

Sagar, B.T.A. (1979). Safe practices for high head outlet gates. *13 ICOLD Congress* New Delhi **Q50**(R27): 459–467.

Sagar, B.T.A. (1995). ASCE hydrogates task committee design guidelines for high-head gates. *Journal Hydraulic Engineering* **121**(12): 845–852.

Sharma, H.R. (1976). Air entrainment in high head gated conduits. *Journal Hydraulics Division* ASCE **102**(HY11): 1629–1646; **103**(HY10): 1254–1255; **103**(HY11): 1365–1366; **103**(HY12): 1486–1493; **104**(HY8): 1200–1202.

Sinniger, R.O. and Hager, W.H. (1989). *Constructions hydrauliques*. Presses Polytechniques Universitaires Romandes, Lausanne (in French).

Task Committee (1973). High head gates and valves in the United States. *Journal Hydraulics Division* ASCE **99**(HY10): 1727–1775.

US Army Corps of Engineers (1957). Sluice entrances flared on four sides. *Hydraulic Design Criteria*, Hydraulic design chart **211**. US Army Engineer Waterways Experiment Station: Vicksburg, Mississippi.

US Army Corps of Engineers (1968). Gate slots – pressure coefficients. *Hydraulic Design Criteria* Hydraulic design chart **212**(1/1). US Army Engineer Waterways Experiment Station: Vicksburg, Mississippi.

Volkart. P.U. and Speerli, J. (1994). Prototype investigation on the high velocity flow in the high head tunnel outlet of the Panix dam. *18 ICOLD Congress* Durban **Q71**(R6): 55–78.

Wagner, W.E. (1967). Glen Canyon diversion tunnel outlets. *Journal Hydraulics Division* ASCE **93**(HY6): 113–134.

Intake vortex at Laufenburg power station on river Rhine (Courtesy B. Etter, VAW)

7

Intake Structures

7.1 INTRODUCTION

The purposes of intake structures are twofold:

1. withdrawal of water for power production, irrigation or drinking-water supply, and

2. inlet of outlet structures.

In the first case, the flow velocity is dictated by the user. Typical design velocities in penstocks or pressure conduits are some metres per second. Accordingly, the water is accelerated to this velocity at the intake structure. A trashrack is provided to retain float and debris.

In the second case, the flow velocity depends on the available pressure head. Intake structures located deeply under the water level thus have a large velocity, and structures close to the water level have velocities as in the first case. For both cases of intake structures, therefore, some similar and some different *principles of design* apply. The common design requirements for all intake structures are:

- insignificant or small setup of vortices for relatively small head on the inlet,

- no separation of flow at the inlet structure, and

- emergency closure.

In addition, *withdrawal structures* have to be protected against float and debris by trashracks, and bottom outlets have a service gate and are designed for diversion of sediment.

Intakes are somewhat contrary to outlets, as regards the hydraulic features. Whereas the latter issue a concentrated jet into the atmosphere or into a tailwater, and energy dissipation is a main concern, intakes behave nearly as a potential flow with the characteristic decrease of pressure due to velocity increase. Depending on whether the head on the intake structure is low or high, significantly different flows are generated, and the associated hydraulic problems are correspondingly different.

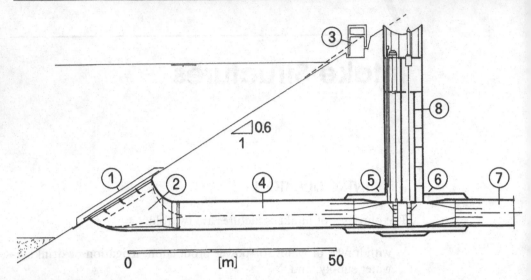

Figure 7.1 Withdrawal structure of Hydroprado power plant (Colombia) with ① intake trashrack, ② intake transition, ③ trashrack cleaner, ④ penstock, ⑤ slide gate, ⑥ wheel gate, ⑦ penstock to power plant, ⑧ gate shaft (Tröndle, 1974)

For *low velocity intakes*, the intake pressure may be so low as to set up an air-water mixture. Further, especially for axisymmetric approach conditions, vortices are generated and a significant swirl component is set up in the intake conduit. A hydraulic design thus involves guidelines to inhibit both a flow with a too strong swirl, and with a too large air content. Otherwise, the conditions in the conduit are poor and significant damage may occur.

For *high velocity intakes*, pressure may locally become so low as to fall below the vapour pressure. Accordingly, the geometry of the intake must be chosen such that pressure is always above the vapour pressure, and cavitation damage is no concern. Based on the trajectory of an orifice jet, the boundary geometry may be defined to inhibit such damages.

Figure 7.1 shows a typical withdrawal structure. The inlet geometry is bellmouth-shaped and is connected with a circular-shaped penstock of 6.15 m diameter. Its capacity is $115 \, m^3 s^{-1}$. The tunnel is lined with 700 mm reinforced concrete. Downstream, 76 m from the intake is the gate shaft, with a slide and a roll gate of $4 \times 5.5 \, m^2$ section each. Both gates are operated from the top of the gate shaft. The transition from the circular penstock to the rectangular gate location is steel-lined. The *trashrack* is connected with the cleaning machine that can be operated also from the gate shaft. Note the *anti-vibration elements* added to the racks, and the complicated transition geometry of the intake. The maximum head on the intake structure is 35 m.

In the following the hydraulic problems related to an intake structure are presented. The hydraulics of high-head intakes are reviewed in 7.2, and those referring to low-head intakes in 7.3. Design relations for the control of cavitation, and vortex flow are particularly addressed.

7.2 BOTTOM INTAKE

7.2.1 Design principles

Inlet structures for bottom outlets have to be safe under all hydraulic conditions. These include: (1) cavitation, (2) vibration and (3) flow stability. In addition, the design of inlet structures should be simple and economic, and the capacity should be large. Because the head on an inlet structure at the bottom of a dam may be well in excess of 100 m, the resulting velocities can be up to $50\,\mathrm{ms}^{-1}$ or even more. These may give rise to low boundary pressure and are a potential danger of *cavitation damage*.

Bottom outlets comprise several parts as shown in Figure 7.2. Depending on the length of the tunnels and the position of the gates relative to the conduit diameter, one may distinguish between the bellmouth-shaped inlet, the pressure tunnel or the *penstock*, the gates and the tailwater tunnel. In particular cases, a pressure tunnel may be missing and the gates are located just downstream of the inlet structure. In the following, attention is paid to the inlet of the intake structure for high head flow. The gate chamber and the tailwater tunnel are described in Chapter 6.

The geometries of the intake portions of withdrawal and bottom outlet structures are similar. Figure 7.3 shows the withdrawal structure of Karakaya dam, Turkey (Özis and Özel 1989). In total, six intakes are connected to separate penstocks, and each has a trashrack of $18 \times 18\,\mathrm{m}^2$ surface area. Together with the bulkhead gates that can be placed with a mobile crane located at the dam crest, a total of 1668 tons of steel was used for the intake structure. The furnishers were Noell (Germany), and Cilting (Turkey).

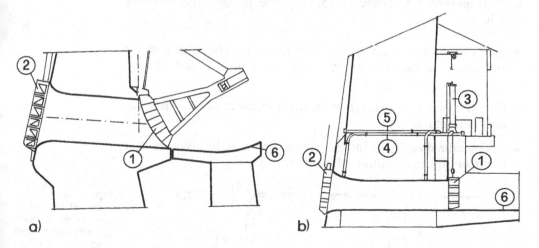

Figure 7.2 Intake portions of Japanese bottom outlets (a) Ayakita dam, (b) Muromaki dam with ① main gate, ② emergency gate, ③ oil-pressure cylinder, ④ pressurized air pipe, ⑤ fill pipe, ⑥ chute (Fujimoto and Takasu, 1979)

Figure 7.3
Upstream view of
Karakaya intake
during construction
(Italstrade S.p.A.,
Milan Italy)

7.2.2 Orifice flow

The definition of the *intake geometry* of a penstock is based on jet flow
from a sharp-crested orifice. For design flow the boundary pressure is
atmospheric as for the corresponding liquid jet. This principle can be
compared with the definition of the crest geometry of the standard
overflow structure, therefore. For corresponding heads, the surface
pressures of both the free liquid jet, and the pressurized flow are
atmospheric, and cavitation is no concern. Based on the vertical jet
flow from an orifice with a horizontal plane and with a standard sharp
crest (Figure 7.4) Liskovec (1955) was able to draw the following
conclusions:

- For orifice diameters in excess of 60 to 70 mm, effects of viscosity
  and surface tension are absent, and scaling follows Froude similarity.

- For an orifice head larger than six orifice diameters, gravity plays an
  insignificant role.

- A generalized equation for the jet geometry may be given, both in
  terms of orifice or contracted diameters.

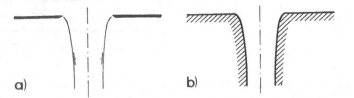

a)   b)

Figure 7.4
Geometry of (a)
axisymmetric liquid
jet with vertical axis,
and
(b) corresponding
intake geometry.

Based on the irrotational flow theory, Kirchhoff and later Von Mises were able to define the jet geometry due to an overpressure in the tank into space by neglecting the effect of gravity (Figure 7.5(a)). Later, Rouse and Abul-Fetouh (1950) defined the geometry of those jets experimentally. For a cylindrical tank of diameter D with an orifice of diameter d, the discharge Q under a head h is

$$Q = C_d A (2gh)^{1/2} \qquad (7.1)$$

with C_d as discharge coefficient and g as gravitational acceleration. Denoting by C_c the *contraction coefficient*, i.e. the ratio of asymptotic jet diameter to the orifice diameter, and by a and A the areas of the orifice and the tank, respectively, one has for potential flow

$$C_d = \frac{C_c}{[1 - (C_c a / A)^2]^{1/2}}. \qquad (7.2)$$

The contraction coefficient C_c depends exclusively on the diameter ratio $\delta = d/D$. For $\delta \to 0$, i.e. a small orifice in a very large tank, the asymptotic value is $C_c = \pi/(\pi + 2) = 0.611$. For the other extreme $\delta \to 1$, one has $C_c = 1$. The relation between C_c and δ is implicitly given by

$$C_c^{-1} = 1 + \frac{2}{\pi} \left[\frac{1}{C_c \delta} - C_c \delta \right] \arctan(C_c \delta). \qquad (7.3)$$

Figure 7.5(b) shows that both C_c and C_d as computed from Eqs.(7.2) and (7.3) are nearly constant if $\delta < 0.20$. For high head intakes, this condition is always fulfilled and the effect of δ is negligible.

Rouse and Abul-Fetouh (1950) determined the jet profiles by using the electrical analogy. It was found that contraction occurs within a distance of nearly $x = d$. Also the pressure distribution on the orifice plate was determined.

For an orifice with a sloping outflow plane, the effect of gravity has to be accounted for. However, due to the short reach along which the jet contracts, this effect can often be neglected, provided the jet Froude number is large as for high head inlets. The details of the jet geometry, and the intake shape for those inlets are presented subsequently.

Figure 7.5
Hydraulic characteristics of orifice flow (a) definition of geometry, (b) (-) contraction and (...) discharge coefficients as a function of diameter ratio $\delta = d/D$

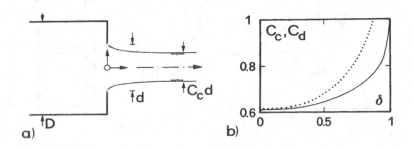

7.2.3 Inlet geometry

The inlet portion of an intake structure has to satisfy various conditions, such as:

- positive boundary pressures;
- absence of cavitation zones;
- continuous decrease of pressure line;
- minor head losses, and thus a good hydraulic performance;
- control of vibration; and
- economic design.

Withdrawal structures must be protected by a *trashrack* against float and debris. Because trashracks may clog, they are usually omitted in bottom outlets. Effects of trashrack *vibrations* are not discussed here because this special topic is thoroughly reviewed by Naudascher and Rockwell (1994). Attention is paid to the *transition curve* from the upstream dam face to the intake penstock. The height of the intake is *a*, and *b* is the width. According to USCE (1957) ellipses or compound ellipses may be used to approximate the inlet geometry. Figure 7.6 shows a definition plot and defines the origin of the coordinate system $(x;y)$. For a sloping dam face, slightly modified curves were presented.

According to USCE, the boundary pressure along the transition may be negative if a single elliptical curve is adopted. Experiments of McCormmach (1968) substantiated this finding and currently, the compound elliptical transition curve is recommended. With the dimensionless coordinates $X = x/a$, or $X = x/b$ depending on whether the vertical or horizontal transition curve is considered, and $Y = y/a$ or $Y = y/b$, respectively, the transition geometry can be defined for a *four sided intake* configuration as

$$(X - 1)^2 + \left(\frac{Y}{0.32} + 1\right)^2 = 1, \quad 0 < X < 0.33; \qquad (7.4)$$

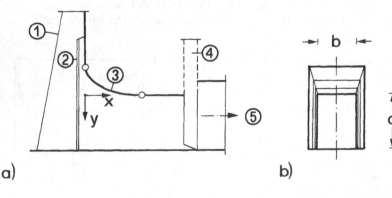

Figure 7.6
Intake geometry of high-head intake (a) longitudinal section, (b) cross section with ① upstream pier face, ② trashrack, ③ transition curve, ④ gate with gate slot, ⑤ penstock

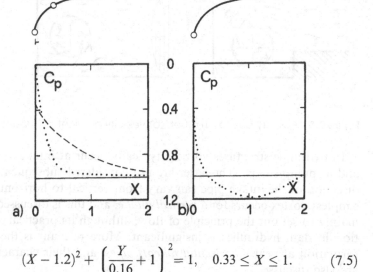

Figure 7.7
Intake transition
curve (a) four sided
and (b) roof curve
only. Transition
geometry (top) and
pressure coefficient
C_p (...) along
vertical plane of
symmetry and top
corner, (---)
horizontal plane of
symmetry (USCE,
1957)

$$(X - 1.2)^2 + \left(\frac{Y}{0.16} + 1\right)^2 = 1, \quad 0.33 \leq X \leq 1. \tag{7.5}$$

Figure 7.7 shows the transition geometry and the *pressure coefficient*

$$C_p = \Delta H / (V^2/2g) \tag{7.6}$$

where $\Delta H(x)$ is the pressure drop from the reservoir into the penstock.
It may be noted that the pressure drop is continuous without any local
minima that could provoke cavitation. Also, the coefficient tends
asymptotically towards $C_p = 1$, i.e. the headloss across the transition
is $\Delta H_t = V^2/2g$ where V is the penstock velocity (Figure 7.7(a)).

For intakes with only a *roof transition* such as shown in Figure 7.6(a)
the USCE (1957) recommended the single elliptical curve

$$(\tfrac{2}{3}X - 1)^2 + (\tfrac{3}{2}Y + 1)^2 = 1, \text{ for } 0 \leq X \leq 1.5, \tag{7.7}$$

where $X = x/a$ and $Y = y/a$. The pressure minimum $C_p = 1.2$
occurs at location $X = 1.2$ (Figure 7.7(b)) but this was considered
acceptable.

7.3 SURFACE INTAKE

7.3.1 Vortex flow

For low submergence, a withdrawal structure can be prone to vortices.
A *vortex* is a coherent structure of rotational flow. It is mainly caused by
the eccentricity of the approach flow to a hydraulic sink, but asymmetric
approach flow conditions and obstruction effects among other reasons
can also set up vortices. Figure 7.8 shows major sources of vorticity.

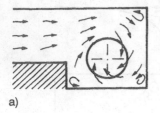 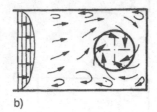 

a)   b)   c)

Figure 7.8 Vorticity due to (a) offset, (b) velocity gradient, (c) obstruction (Knauss, 1987)

In hydraulic structures vortices typically occur at intake structures and at pump sumps. The latter are not considered subsequently. The direction of the intake pipe can vary from vertical to horizontal. The simplest vortex occurs for a *vertical intake*, and this is discussed below mainly to set out the principle of flow, although its practical application in dam hydraulics is insignificant. More relevant is the vortex formation at nearly *horizontal takeoff pipes*, and their characteristics are also discussed.

Vortices have four main disadvantages in hydraulic designs:

- *air entrainment*, with effects on hydraulic machinery or hydropneumatics, and pressure surges,

- *swirl entrainment*, with increase of headloss and reduction of efficiency in hydraulic machinery,

- *enhancement of cavitation and vibration* with a reduced longevity of important mechanical parts, and

- *entrainment of floating material* such as wood or ice, and blockage of screens, or damage of coatings.

Ideally, the transition from free surface to pressurized flow should be uniform, steady and of single phase. Air entrainment is the worst consequence of an intake structure, and an acceptable hydraulic design has to avoid this condition in particular.

7.3.2 Vertical intake vortex

The basic type of vortex flow occurs in a large cylindrical containment with a much smaller circular sharp-crested orifice in its centre. But even for such a simple configuration, no physical solution is currently available. Effects such as streamline curvature and turbulence are significant, not to mention viscosity and surface tension for scale models, on which most of the available data were obtained.

The *Rankine combined vortex* is a simplified model of the typical vortices that occur in nature (Figure 7.9). It is composed of a *forced*

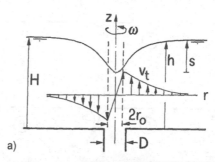

Figure 7.9
Rankine vortex
(a) free surface and
distribution of
tangential velocity
(b) typical flow
pattern

vortex in the core zone and a *free vortex region* outward from the core. In the core zone, viscosity affects the flow and yields zero central velocity, whereas potential flow dominates the free vortex zone. The two flow regions are matched at radius r_o which is a fraction of the orifice radius $R = D/2$.

In the vortex core, the tangential velocity v_t is assumed to vary linearly with the radial coordinate r and the angular velocity ω as

$$v_t = \omega r, \quad 0 < r \leq r_o. \tag{7.8}$$

In the free vortex region, the circulation Γ is constant, and one has

$$v_t = \frac{\Gamma}{2\pi r}, \quad r_o < r. \tag{7.9}$$

The simplified model does not account for vertical velocities nor for vertical variations of the horizontal velocities.

According to Anwar (1966), the depth s of the *free surface depression* can be approximated as

$$s = 0.60[v_{tM}^2/(2g)] \tag{7.10}$$

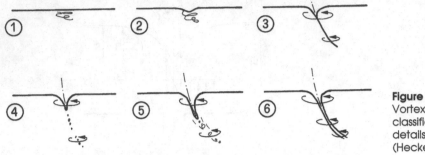

Figure 7.10
Vortex type
classification,
details see text
(Hecker, 1984)

with $v_{tM} = v_t(r = r_o)$ as maximum tangential velocity. The location of maximum velocity was found to be nearly $r_o = (3/4)R$. Accordingly, the relation between ω and Γ is $\Gamma/\omega = 2\pi r_o^2 = 3.5R^2$.

Vortices can be classified according to their strength intensity. It was found that a visual *classification* is superior to defining the circulation due to difficulties in measuring a swirl parameter. Figure 7.10 is based on this concept and six *vortex strengths* are distinguished:

1. coherent surface swirl,

2. surface dimple, with a coherent surface swirl,

3. dye core to intake, with a coherent swirl along column,

4. vortex pulling floating trash, but no air,

5. vortex pulling air bubbles to intake, and

6. vortex with full air core.

A basic vortex, sometimes also referred to as *bathtub vortex*, depends on various dimensionless parameters, including:

$$\text{submergence Froude number} \quad \mathbf{F} = \frac{Q}{(\pi/4)D^2(gh)^{1/2}}$$

$$\text{circulation number} \quad \mathbf{N}_\Gamma = \frac{\Gamma D}{Q}$$

$$\text{radial Reynolds number} \quad \mathbf{R} = \frac{Q}{\nu h}$$

$$\text{Weber number} \quad \mathbf{W} = \frac{Q}{(\pi/4)D^2(\rho D/\sigma)^{1/2}}$$

where Q is discharge, D pipe diameter, h intake submergence, ν kinematic viscosity, ρ mass density and σ surface tension. The relative submergence $y = h/D$ is also of importance. For $\mathbf{W}^2 > 10^4$, the effect of surface tension is insignificant (Hecker, 1987). Also, viscous effects are dominant mainly for $\mathbf{R} < 3 \times 10^4$. For prototype structures, the vortex characteristics are thus determined by the parameters $\mathbf{F}$, $\mathbf{N}_\Gamma$ and y. The difficult parameter to estimate is the circulation Γ, because it may even be interrelated to the other parameters. Usually, Γ can be estimated only on scale models or prototype structures if available, or from complex numerical simulations. For more details see Hecker (1987). Scale effects which are particularly important for the prediction of air entrainment in prototype structures were addressed by Chang and Prosser (1987).

7.3.3 Limit submergence

To avoid air entrainment and counter swirl entrainment in an intake, an adequate submergence is required (Jain, et al., 1978). Figure 7.11 refers to the three basic configurations of intakes that can be characterised with the *orientation* Φ relative to the vertical direction, and the *distance* of the intake centre from the invert. The orientations in Figure 7.11 are $\Phi = 0$, $\pi/2$, and π, respectively.

The *limit submergence height* h_L was defined by Knauss (1987) as the height under which the vortex type 5 is generated. For $h \geq h_L$ air bubbles are entrained continuously but intermittently, and there is no air core into the intake structure. According to Knauss, the effects of circulation and Froude number can be combined in a *swirl number* $\mathbf{C} = \Gamma/(gD^3)^{1/2}$ where D is the pipe diameter. The limit submergence varies thus with

$$h_L/D = f(\Phi, \mathbf{C}).\qquad(7.11)$$

The limit submergence was evaluated for various type structures, such as those shown in Figure 7.11. It was found that the relation between the limit submergence and $\mathbf{C}$ is nearly linear. One may thus write with C as constant of proportionality

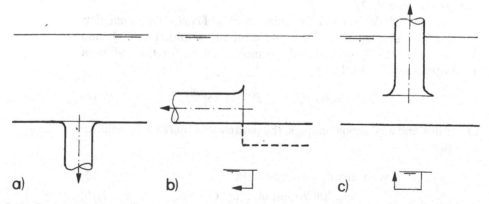

Figure 7.11 Basic intake types and notation (a) vertical intake pipe, (b) horizontal intake pipe, (c) pump intake. Intake orientation Φ.

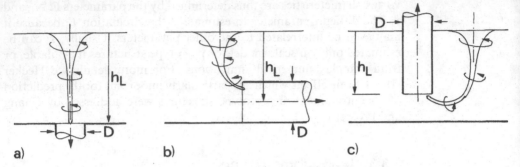

Figure 7.12 Limit submergence height h_L for (a) vertical intake, (b) horizontal intake, (c) pump intake.

$$h_L/D = C\frac{\Gamma}{(gD^3)^{1/2}}. \qquad (7.12)$$

For the same swirl number, the effect of structural geometry may thus be described with the constant C. It can further be demonstrated that a simple relation exists between the radius r_o as defined in Figure 7.9 and the maximum tangential velocity. With $C = C(\Phi)$, i.e. as a function of orientation only, the limit submergence height is

$$h_L/D = [C(r_o/D)]^2. \qquad (7.13)$$

This equation provides a mean to determine the constant C.

Figure 7.12 refers to the typical structures shown in Figure 7.11. Because of the intake shape, the flow has to contract at the inlet section. If a vortex has a central air core extending beyond the inlet section, air is continuously entrained and the intake submergence is insufficient. The indicators of intake direction and swirl number are the *surface vortex diameter* and the *vortex length* up to the inlet section. Both parameters increase as the intake direction and the swirl number increase (Figure 7.12).

Knauss (1987) determined the main features of *horizontal intake flow* under limit submergence. From the data of Amphlett (1976), and Anwar, et al. (1978) he deduced a simple relation for the radius of maximum tangential velocity

$$r_o/D = 0.0109[(h_L/D) + 1.45]^{1/2}. \qquad (7.14)$$

From this and analogous findings, the parameter C may be determined as follows

$$\text{vertically downward intake} \quad C = 110, \qquad (7.15)$$
$$\text{horizontal intake} \quad C = 90, \qquad (7.16)$$
$$\text{vertically upward intake} \quad C = 75. \qquad (7.17)$$

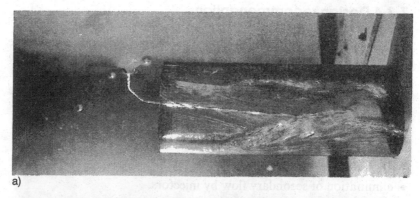

a)

b)

Figure 7.13
Inflow to a
horizontal pipe,
(a) Vortex intensity
6, (b) swirling pipe
flow due to intake
vortex

The *effect of orientation* is thus considerable and yields an increase or decrease of 20%, respectively, compared to the horizontal intake arrangement. For oblique intake orientation, one may linearly interpolate between the numbers previously specified.

Based on the tests of Gordon (1970) and Hecker (1981), Knauss (1987) recommended for the *minimum submergence* of intakes with normal approach conditions

$$(h_L/D) = 1 \text{ to } 1.5, \qquad \text{for } \mathbf{F} \le 0.25, \tag{7.18}$$
$$(h_L/D) = (1/2) + 2\mathbf{F}, \quad \text{for } \mathbf{F} > 0.25, \tag{7.19}$$

with $\mathbf{F} = V/(gD)^{1/2}$ as the *pipe Froude number*. For $\mathbf{F} < 0.25$ the intake structure is large and a typical pipe velocity is $2\,\text{ms}^{-1}$. For $\mathbf{F} > 2$ a small or medium intake structure prevails, with a typical pipe velocity of $4\,\text{ms}^{-1}$. These recommendations do not include special features of vortex suppression.

Figure 7.13 shows inflows to a horizontal intake pipe. In Figure 7.13(a), the full air core vortex can be seen, and the vorticity set up in the pipe flow can be seen in Figure 7.13(b).

7.3.4 Design recommendations

The designer of an intake structure has various methods to improve the vortex characteristics. These refer mainly to the approach flow geometry, and special vortex suppression devices.

According to Rutschmann, et al. (1987) the *approach flow* can be modified by:

- uniformation of flow by appropriate elements,
- elements that direct the flow to the intake,
- elimination of secondary flow by injectors,
- streamlining of boundaries and piers,
- partial gate closure, and
- acceleration of approach flow by a tapering section.

Also, improvement can be obtained by *elongation of streamlines* towards the intake:

- higher tailwater submergence,
- lower intake level,
- changing the inflow direction,
- horizontal roof above intake, and
- reduction of the approach velocity by widening the cross-section.

Anti-vortex devices that are frequently used for vortex suppression are:

- vertical rows of walls,
- horizontal beams,
- floating rafts, and
- flow straighteners.

All these devices, and lots of other propositions aim at breaking up a significant swirl portion (Gulliver, et al., 1986). It should be noted that the devices induce a *self-perturbation*, and that they should be used only after detailed model observation in order to inhibit overforcing. Because the mechanics of vortices is complex, physical modelling is recommended for all cases where a failure can lead to a significant loss.

Figure 7.14 is intended to review some successful techniques adopted in existing intake structures (Rutschmann, et al., 1987). In Figure 7.14(a) the vortex is suppressed with buoys and a horizontal beam.

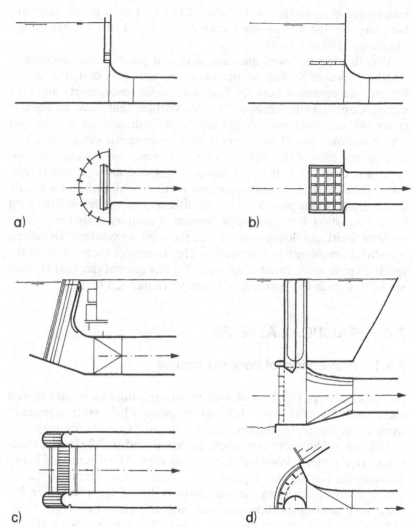

Figure 7.14
Intake structures of
(a) Gevelinghausen
(Germany), (b) Mt.
Elbert (USA), (c)
Bremgarten
(Switzerland), (d) El
Cajon (Honduras)

The maximum intake head is nearly 5 m, the minimum intake section 4.5 m × 11.4 m and the design discharge 92 m^{3}s^{-1}.

The intake of Mt. Elbert powerplant (USA) is protected by a horizontal rack against vortices (Figure 7.14(b)). The intake head may vary between 10 and 20 m, the intake diameter is 4.75 m and the design discharge is 102 m^{3}s^{-1}.

Bremgarten powerplant (Switzerland) has an injector that favours parallel flow to the intake pipe, and the stagnant surface water is suppressed (Figure 7.14(c)). Also, floating debris is forced to move towards the trash rack. The intake head is 15.5 m, the intake section 8.6 m^2 and the design discharge 100m^{3}s^{-1}.

For El Cajon (Honduras) an intake tower stabilizes the vortex formation and reduces rotation by sucking in water from the dead

water zone close to the dam (Figure 7.14(d)). Intake heads may vary
between 9 m and 41.5 m, the intake diameter is 12 m, and the design
discharge $1780\,\text{m}^3\text{s}^{-1}$.

After having reviewed the current design practice, the question is
whether vortices should be prevented or not. The design policy is
towards an optimum between *hydraulic performance*, safety and economy. Consequently, there are lots of vortices that have no adverse
effect on the performance and safety of hydraulic structures, and
whose removal would result in a highly uneconomic design. If an air
core of, say 10%, of the intake diameter is considered typical, its cross-
section is only 1% of the pipe section. A typical air ingestion is then
lower than 2% even for strong air core vortices. Unless the air is locally
concentrated in the pipe system, little difficulty should result from such
low air ingestion. Pipe bends downstream of the inlet section may cause
as much swirl and flow asymmetry as these intake vortices. Therefore,
one should evaluate whether the flow characteristics are adverse for the
specific application, rather than adopting as a general rule that vortices
are to be avoided as a matter of principle (Knauss, 1987).

7.4 PRACTICAL ASPECTS

7.4.1 Prevention of floating matter

Intake structures are covered with water, and floating matter cannot
enter easily the pipe inlet. However, suspended load such as roots of
trees or ingestion of floating matter by surface vortices may give
significant problems of operation, as mentioned in 7.2. Intake structures are therefore protected with a *trash rack*. The distance of bars e
depends on the degree of protection. As a rule of thumb, $e=(2/3)E$
where E corresponds to the minimum width of the plant. The bar
distance is sometimes chosen so that large fish cannot enter.

For outlet structures used for *flood diversion*, racks should be
omitted. Accordingly, the dimension of the intake has to be chosen
so that obstruction is impossible even from the largest floating matter.
For bottom outlets, a large bar trash rack is normally provided to
retain logs of wood contained in the sediment, thus, preventing clogging of the outlet gates.

For withdrawal structures, the velocity is typically between $0.5\,\text{ms}^{-1}$
and $1\,\text{ms}^{-1}$, over periods of hours or days. Trash racks have thus to be
designed against flow-induced vibration. These are enhanced by
separation of flow from the bars. Figure 7.15 relates to the *Strouhal
number* $S=fd/V$ for various bar profiles, from which the natural
frequencies may be determined. For bars of $d=100\,\text{mm}$ diameter
and $V=1\,\text{ms}^{-1}$ flow velocity, these are typically between 0.3 and 1
Hz. Consequently, the bars have to be designed and attached so that
their natural frequency f is increased by, about 20%. These indications

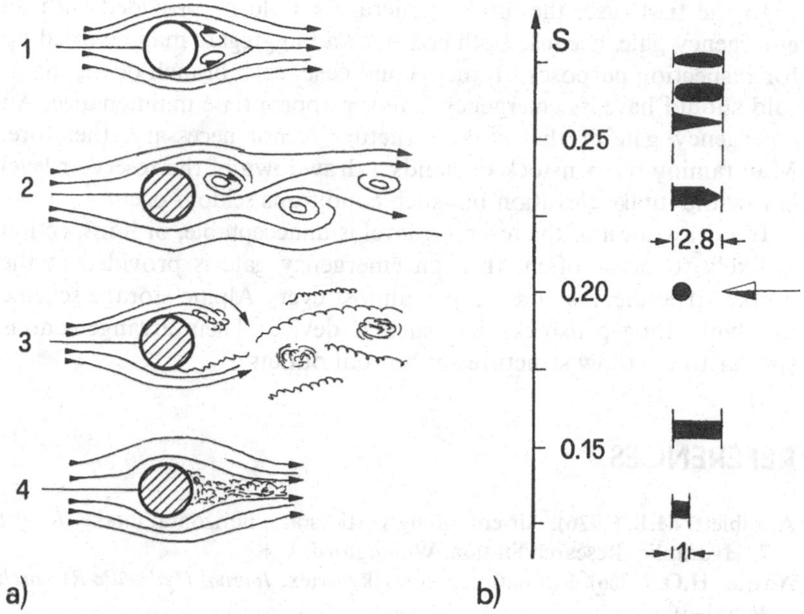

Figure 7.15
Natural frequencies
of typical bar
profile. (a) Flow
conditions and
vortex shedding for
a circular-shaped
cylinder with
$R = Vd/\nu = $ (1) 20, (2)
10^2, (3) 10^4 and (4)
10^8. (b) Strouhal
numbers **S** for
various rack shapes
of width d and
length d, or $2.8d$
(Levin, 1967).

relate also to rack tables and the suspensions. Details on the design are extensively discussed by Naudascher and Rockwell (1994).

The design of a trash rack contains means for cleaning. For trash racks close to the free surface of the reservoir, a *rack cleaner* is recommended. For trash racks of intakes located deeply below the free surface a rack pedestal should at least be considered in the design, that is accessible during drawdown of the reservoir level. Cleaning is then possible with racks or other suitable tools. Clearly, trash racks with a small bar distance clog faster than if the bar distance is wide. Therefore, and because of a reduction of head losses, the bar distance is not reduced beyond the minimum value previously mentioned.

7.4.2 Emergency gate closure

The possibility of placing emergency gates in front of a regulating gate has already been discussed in Chapter 2 and section 6.2. In the following, emergency gates of withdrawal structures are considered.

The discharge of withdrawal structures for water use is regulated close to the user, and not at the intake structure. It may be a pump at the pumping station, a turbine at the powerhouse, or valves for irrigation works. Two cases may be distinguished:

1. Pressure conduits connected to a single aggregate, or

2. Pressure conduits discharging into a set of parallel aggregates and thus terminating in a manifold pipe.

In the first case, the intake structure should be provided with an emergency gate, because both conduit and aggregate may be dried up for inspection purposes. In the second case, each branch of the manifold should have its emergency gate for appropriate maintenance. An emergency gate at the intake structure is not necessary, therefore. Maintaining the penstock demands a drawdown of the reservoir level below the intake elevation but such conditions seldom occur.

If a drawdown of the reservoir level is unacceptable, or if inspection is likely to occur often, then an emergency gate is provided at the intake structure. As an example, almost every Alpine storage scheme involving long penstocks has such a device. Their arrangement is similar to overflow structures of bottom outlets.

REFERENCES

Amphlett, M.B. (1976). Air-entraining vortices at a horizontal intake. *Report 7*. Hydraulics Research Station: Wallingford, U.K.

Anwar, H.O. (1966). Formation of a weak vortex. *Journal Hydraulic Research* **4**(1): 1–16.

Anwar, H.O., Weller, J.A., Amphlett, M.B. (1978). Similarity of free vortex at horizontal intake. *Journal Hydraulic Research* **16**(2): 95–105.

Chang, E. and Prosser, M.J. (1987). Basic results of theoretical and experimental results. Swirling flow problems at intakes: 39–55. *IAHR Hydraulic Structures Design Manual* **1**. J. Knauss, ed., Balkema, Rotterdam.

Fujimoto, S. and Takasu, S. (1979). Historical development of large capacity outlets for flood control in Japan. *13 ICOLD Congress* New Delhi **Q50**(R25): 417–438.

Gordon, J.L. (1970). Vortices at intakes. *Water Power* **22**(4): 137–138.

Gulliver, J.S., Rindels, A.J., Lindblom, K.C. (1986). Designing intakes to avoid free-surface vortices. *Water Power and Dam Construction* **38**(9): 24–28.

Hecker, G.A. (1981). Model-prototype comparison of free surface vortices. *Journal Hydraulics Division* ASCE **107**(HY10): 1243–1259; **108**(HY11): 1409–1420; **109**(3): 487–489.

Hecker, G.A. (1984). Scale effects in modelling vortices. *Symposium Scale Effects in Modelling Hydraulic Structures* Esslingen, H. Kobus, ed., **6**(1): 1–9.

Hecker, G.A. (1987). Fundamentals in vortex intake flow. Swirling flow problems at intakes: 13–38. *IAHR Hydraulic Structures Design Manual* **1**, Balkema, Rotterdam.

Jain, A.K., Ranga Raju, K.G., Garde, R.J. (1978). Vortex formation at vertical pipe inlet. *Journal Hydraulics Division* **104**(HY10): 1429–1445; **105**(HY10): 1328–1336; **106**(HY1): 211–213; **106**(HY9): 1528–1530.

Knauss, J. (1987). Swirling flow problems at intakes. *IAHR Hydraulic Structures Design Manual* **1**, Balkema, Rotterdam.

Levin, L. (1967). Problèmes de perte de charge et stabilité des grilles de prises d'eau (Problems of head loss and stability of intake racks). *La Houille Blanche* **22**(3): 271–278 (in French).

Liskovec, L. (1955). Suitable inlet form of pressure conduits. *6 IAHR Congress*, The Hague **C**(14): 1–11.

McCormmach, A.L. (1968). Dworshak dam spillway and outlets hydraulic design. *Journal Hydraulics Division* ASCE **94**(HY4): 1051–1072.

Naudascher, E. and Rockwell, D. (1994). Flow-induced vibrations. *IAHR Hydraulic Structures Design Manual* 7, Balkema, Rotterdam.

Özis, Ü. and Özel, I. (1989). Karakaya dam and powerplant. *Journal Water Power & Dam Construction* **41**(7): 20–24.

Rouse, H. and Abul-Fetouh, A.-H. (1950). Characteristics of irrotational flow through axially symmetric orifices. *Journal Applied Mechanics* **17**(12): 421–426.

Rutschmann, P., Volkart, P., Vischer, D.L. (1987). Design recommendations. Swirling flow problems at intakes: 91–100. *IAHR Hydraulic Structures Design Manual* 1, J. Knauss ed., Balkema, Rotterdam.

Tröndle, E. (1974). Hidroprado / Columbien Wasserkraftanlage. *Wasserwirtschaft* **64**(2): 33–41 (in German).

USCE (1957). Sluice entrances. *Hydraulic Design Charts* **211**(1), and **211**(1/1), revised 1964. United States Army Corps of Engineers, Waterways Experiment Station: Vicksburg, MI.

a)

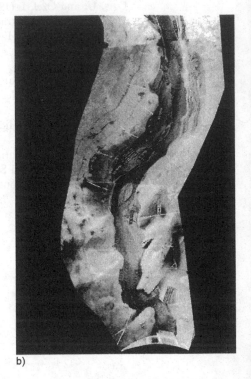
b)

Gebidem dam, Switzerland, Flushing of sediment. (a) Orifice flushing at bottom outlet (VAW 3872) (b) Reservoir flushing (VAW 5741). Direction of flow from top to bottom (*ICOLD* **Q54**, R25)

8

Reservoir Sedimentation

8.1 INTRODUCTION

Sedimentation of a reservoir reduces the storage capacity. According to ICOLD (1988), an average loss of capacity amounts to 1% per year, when the existing storage volumes of all dams worldwide are summed up. This loss is considerable and reservoir sedimentation is one of the primary problems to be dealt with in the 21st century. Reservoir sedimentation is thus of great relevance to the management of water resources. According to Graf (1983) we are currently at an early stage of predicting the deposition pattern in reservoirs, and much effort has to be done in the future.

Basically, two methods to reduce reservoir sedimentation are available:

1. Sediment retention in the catchment area, and

2. Sediment removal from the reservoir.

The *retention of sediments* in the catchment area deals with the root of the problem. In a large catchment, this is hardly possible due to economic reasons. In a small or medium catchment area, the *erosion process* can be reduced with two methods:

- soil conservation, and

- measures in the water network.

Methods that involve *soil conservation* are lengthy and costly. They become effective only after decades. Soil conservation may be applied only if it is supported by agricultural development. In non-cultivated areas such activities are practically disregarded.

Methods that involve measures in the *water network* are based on the elimination of sediment concentrations. Accordingly, slide areas along rivers are prevented by river works, and river erosion both in depth and in width is inhibited by check dams. Sediment retention basins may be arranged at particular locations, but these have to be cleaned periodically.

All previous measures apply to sediments that are bed load. *Suspended load* cannot be retained in a catchment, and this reaches the reservoir in any case. In the following, questions relating to both the transport of suspended load into a reservoir, and the sedimentation processes are dealt with.

The inflow to a reservoir contains, besides water, both suspended and bed load. The amount of *suspended load* is often much larger than the *bed load*. At the river mouth into the reservoir, the bed load settles and contributes to the delta development. The grains of the suspended load settle close to the delta, and finer particles are taken into the reservoir where they settle mainly at stagnation-water zones. Extremely fine sediments remain suspended and leave the reservoir with water again.

The question relating to the content of various sediment sizes depends on the reservoir. A typical measure is the ratio between the storage volume and the annual sediment transport. The larger this ratio is, the better is the retention potential of the reservoir to catch all sediments. For a small ratio, the reservoir degrades to a storage basin for gravel as previously mentioned under the measures against erosion.

The inflow to a reservoir is a mixture of water and sediment. Due to the difference in density, so-called *density currents* develop from the mixture of water and fine sediments. Density currents in general are two-phase flows with small density differences, such as cold and warm water, water and sludge but not air-water flows, where the density ratio is nearly 1000. Also, the two fluids should be miscible and the density difference be a function of differences in the temperature, the salt content or the sediment content, independent of pressure and elasticity of the fluids involved.

The sediment laden flow has a larger density than the reservoir water and moves along the reservoir bottom towards the dam. Figure 8.1 shows a typical flow configuration with A the river inflow, B the clear water reservoir, and C the recirculation zone. The density current is seen to move along the talweg towards the dam outlet. Upstream the coarse material is deposited as a delta and from the plunge point, the current dives along the reservoir bottom to reach the outlet zone. Upstream from the outlet crest, a deposition of fine material develops in the upstream direction.

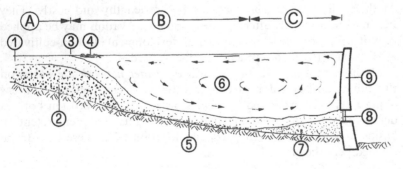

Figure 8.1
Density current in a reservoir with ① approach flow, ② delta, ③ plunging point, ④ floating debris, ⑤ density current, ⑥ clear water, ⑦ sediment deposition, ⑧ outlet, and ⑨ dam.

Due to local sediment deposition, the density current gets diluted along the trajectory. In the delta, where the velocity of the density current is already reduced, particles larger than, say, 0.05 mm are deposited. The finest silt particles with a diameter less than 0.01 mm are carried up to the outlet, while the intermediate fraction is deposited in zone B.

8.2 METHODS OF SEDIMENT REMOVAL

8.2.1 Direct methods

The removal of sediments from a reservoir is linked to questions such as: (1) which sediments are concerned? and (2) how should these be removed? The answers can be given as follows:

- The river delta grows into the top layer of the reservoir. Consequently the *usable volume* of the reservoir is progressively reduced.

- The density currents fill the lowest layers of the reservoir and thus use the *dead volume* of the reservoir.

- If the toe of the delta reaches the bottom outlet, or if the density current has risen above its intake elevation, then the functioning of the bottom outlet is reduced. If this process continues then the intake structure is also covered with sediments, and becomes useless.

The measures to remove sediments from a reservoir are thus directed to the conservation of either the usable volume, or the functioning of the bottom outlet, or even both.

In the following, the direct measures that may be applied to a reservoir are presented. Various methods have been proposed, namely:

- gravel basin,
- front basin with or without sediment bypasses,
- reservoir dredging, and
- sediment releases.

8.2.2 Gravel basins

The gravel basin is located just upstream from the reservoir and can be designed as a *storage zone*. It must normally be cleared annually or after each large flood. There results a gravel deficit in the downstream river reach, however, which may lead to degradation of the river bed. This erosion process, in turn, may reduce the effect of the gravel basin (Figure 8.2).

A gravel basin is therefore effective only if the:

- tailwater river is resistant against large scale-erosion,

- erosion is reduced by suitable protection, or

- gravel basin is located immediately upstream from the reservoir (see 8.2.3).

Figure 8.2
Gravel basin at grade break upstream from the reservoir, with (– – –) possible erosion of tailwater river due to sediment deficit

8.2.3   Front basin

To reduce sedimentation in a reservoir, a *secondary dam* may be erected at the reservoir inlet zone that retains the bed load. Usually this secondary dam is fully submerged under maximum reservoir level, and gravel as well as sand may effectively be settled. The silt fraction is carried into the reservoir and deposits (Figure 8.3).

To prevent sedimentation of the gravel retention dam, it must be cleared periodically, either by removing the sand and gravel or by flushing it into the river beyond the main dam. The bypass is in operation only during floods and its tractive forces must be so large as to prevent clogging by larger deposits of granular material. The bypass must further resist the abrasion (see below). The inlet to the bypass must satisfy the conditions as stated in Chapter 3.

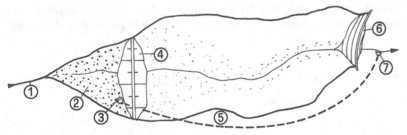

Figure 8.3
Gravel storage dam to counter reservoir sedimentation
① approach river,
② pre-reservoir,
③ flush inlet
④ secondary dam,
⑤ flush bypass,
⑥ main dam,
⑦ flush outlet

8.2.4   Reservoir dredging

Depending on the local depth, the quality of sediments, and the transport systems, sediments may be removed either from the shore or from a boat. Because the cohesion of the sediments settled is large, simple suction devices cannot be used. Therefore, the head of such a device has to be extended with water jet nozzles or with a rotating head to loosen the material for easy dredging.

A particular solution was proposed for small lakes in Italy (Roveri, 1981). In this method a dredger mounted on a swimming platform is connected to a pipeline that discharges at an orifice outlet of the dam. The *hydrotransport* is effected by the siphon alone and the effect of the conventional sediment outlet is extended into the reservoir. The dredger has thus no pumps because the head between the reservoir level and the outlet is sufficient for hydrotransport (Figure 8.4).

If the reservoir inlet zone is relatively narrow and the gravel delta is well defined, then a *dredging system* may be effective provided the reservoir level is virtually constant (Figure 8.5). The removal system is either mobile or is operated from the shore. Such a dredging system has proven effective and simple. The material removed can be used as dam filler or for concrete. However, when containing too much organic matter or sand and sludge, it has to be piled close to the removal location. Further details are presented by Breusers, et al. (1982). It should be mentioned that dredged sludges are not easily deposited because of dewatering and quality. Therefore, the material is often returned to the tailwater river.

Figure 8.4
Sediment removal with a dredger connected to syphon pipe with ① head of pipe, ② dredging platform, ③ connecting hose, ④ buoys for pipeline suspension, ⑤ bottom outlet pipe, ⑥ tailwater

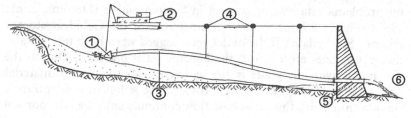

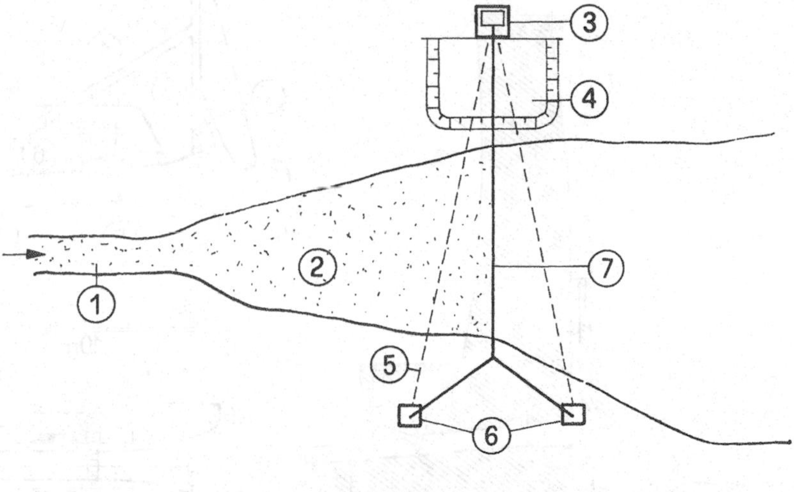

Figure 8.5
Permanent dragline dredging station at reservoir inlet zone ① approach river, ② delta, ③ dredging tower, ④ sediment deposit zone, ⑤ stabilization cable, ⑥ fixation for cable, ⑦ main cable

8.2.5 Clearing of dam outlet

The bottom outlet of a dam is located normally in the front portion of the reservoir sediments. Therefore, the outlet is used to clear the

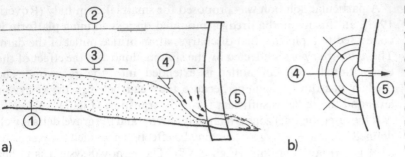

Figure 8.6
Flushing of sediment
front portion by
bottom outlet,
(a) section, (b) plan
view with
① reservoir deposits,
② maximum and
③ minimum
reservoir levels,
④ flushing cone,
⑤ bottom outlet

material settled just upstream. Flushing of material leaves a cone with
side slopes up to 1:1. The effect of the outlet is thus highly reduced (Figure
8.6), and only a small portion of the entire reservoir may be flushed.

Because the bottom outlet is not permanently in operation, it may
clog due to the advancement of the sediment front or due to under-
water slides. *Clogging* may also occur due to logs, but this is not
typical. An upper sediment concentration of 5% is reported to yield
no problems with hydrotransport in bottom outlets (Dawans, et al.,
1982). Figure 8.7 shows the bottom outlet of the Gebidem storage
scheme, Switzerland. If the outlet gets clogged when both the flap and
the sector gates are opened, the injector pit is operated. Due to the
enormous hydraulic grade in the sediment front, packets of material
pass as a water-sediment flow. Provided the sediment concentration
stays within limits, this two-phase flow continues until the inlet portion

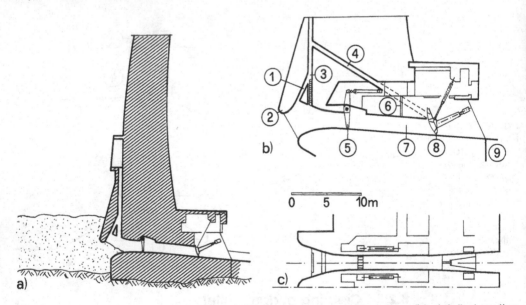

Figure 8.7 Gebidem bottom outlet with flushing gates (a) overall section, (b) detailed section,
(c) plan. ① Jetting pipe, ② guard nose, ③ bulkhead gate, ④ compensation water pipe, ⑤ check
valve, ⑥ by-pass, ⑦ steel-blinded bottom outlet, ⑧ radial gate, ⑨ protected tailwater invert.
(Dawans, et al. 1982)

to the bottom outlet is cleared. Otherwise, the injector has to be operated again to clean the outlet.

The design of the *injector pit* has to be based on the following requirements:

- the inlet to the pit should not self-clog,
- the inlet level is under the minimum reservoir level,
- the efficiency is sufficient to clear the bottom outlet, and
- the structure is as small as possible not to interfere with the normal outlet.

Note that clogging is most probable for bottom outlets without a special anticlogging device.

8.2.6 Reservoir emptying

The effect of outlet clearing is local, and counters reservoir sedimentation only for small basins. Emptying the entire reservoir is a standard procedure that involves the *flushing* of sediments into the tailwater. The process may be compared with extreme erosion of a river having a sandy or silty bed. Figure 8.8 shows the temporal development of the flushing and the resulting reservoir erosion down to the unerodable bottom. Erosion starts at the cone close to the outlet and develops upstream along the talweg. The erosion process ends when the tractive forces become smaller than the resistive forces, and an equilibrium bottom topography has developed. Note the canyon-like erosion along the talweg, and the relatively small erosion along the reservoir flanks mainly due to bank slides.

Flushing of a reservoir as previously described is more efficient than operating the bottom outlet under nearly full reservoir level. However, the dam gets out of operation, and large quantities of water are lost. The capacity of the bottom outlet has an effect on the success of

Figure 8.8
Flushing a reservoir (a) streamwise section, (b) typical section with ① remaining sedimentation, ② erosion cone, ③ maximum reservoir level, ④ reservoir bottom prior to flushing, and ⑤ after some flushing time, ⑥ minimum reservoir level, ⑦ bottom opening

flushing. For too small an outlet, the flushing is only local, whereas artificial *tailwater floods* are generated under too large flushing discharges. An optimum *flushing programme* has to be determined for each reservoir with a maximum effect in the reservoir and a minimum

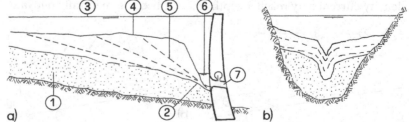

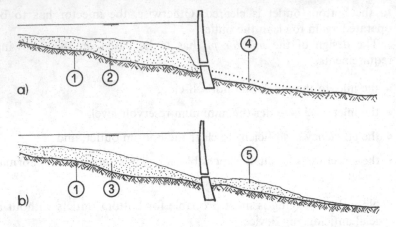

Figure 8.9
Effect of (a) reservoir sedimentation and (b) reservoir flushing on tailwater river, ① original river bed, ② sedimentation zone, ③ reduced zone after flushing, ④ degradation, ⑤ tendency of aggradation of tailwater invert

impact on the tailwater. Accordingly, the transport capacity of the downstream river should be so large that deposits are small (Figure 8.9). Environmental concerns regarding fish and water flora should be accounted for to respect the ecological balance.

8.3 DENSITY CURRENTS

8.3.1 Definition

A density current is a layer of fluid flowing in an ambient fluid of different density. The density difference may result from differences in temperature, dissolved solid concentration, salinity or due to different fluids involved. Depending on the density ratio of the inflow and the reservoir flow, a buoyant or a plunging density current may form, and the latter configuration is considered in what follows.

Density currents have features in common with open channel flows but differ essentially because the *buoyancy* of the surrounding fluid reduces the gravity force. Figure 8.10 shows the zones involved in a density current. The approach flow ① of density ρ_o may be analysed with conventional hydraulics. At the plunge point ② equilibrium between the density difference and the baroclinic pressure exists (see below). Beyond the plunge point a two layer flow ③ develops associated with mixing across the *interface*. If the reservoir is stratified the density current may reach a depth where it becomes neutrally buoyant.

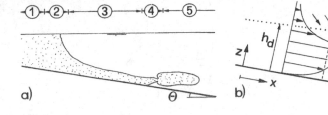

Figure 8.10
Density flow into stratified reservoir (a) flow zones, (b) notation with (...) interface.

Thus, separation ④ from the inclined bottom occurs to form an intrusion ⑤.

The flow of a density current depends mainly on the *reduced gravity* $g' = (\Delta\rho/\rho_o)g$ where g is the gravitational acceleration, ρ_o the approach density and $\Delta\rho$ the density difference between the inflow and the ambient fluid. Also, provided that the Reynolds number $\mathbf{R} = uh/v$ with u and h as the representative streamwise velocity and flow depth and v as the kinematic viscosity is in excess of, say 2×10^3, the current is turbulent with *mixing* along the interface. The stability of the interface depends on the *Richardson number* (Alavian, et al., 1992)

$$\mathbf{Ri} = \frac{g'h\cos\theta}{u^2} \tag{8.1}$$

where θ is the slope of the incline, and $\mathbf{Ri}$ expresses the balance between the effective gravity perpendicular to the interface and the shear across the interface. Decreasing $\mathbf{Ri}$ increases the amount of mixing. The buoyancy may thus either accelerate and destabilize, or damp and stabilize a density current.

8.3.2 Plunge point

With $\mathbf{F}_p = V_o/(gh_p)^{1/2}$ as the Froude number based on the upstream velocity V_o and the depth h_p at the plunge point, Q_o as approach discharge, and b as the approach width, one may write for $0.1 < \mathbf{F}_p < 0.7$ in terms of 1D hydraulics

$$h_p = \left(\frac{Q_o^2}{g'b^2}\right)^{1/3} \mathbf{F}_p^{-2/3}. \tag{8.2}$$

For three-dimensional inflow geometry, i.e. sloping bottom and expanding sides, the plunging characteristics are more complex, involving stalled and no-stalled flows. Some generalized relations are also provided by Graf (1983) and Alavian, et al. (1992).

8.3.3 Equilibrium flow

The *uniform flow* of a density current (zone in Figure 8.10(a)) depends on the bottom topography and the stratification characteristics. Also, the Richardson number of the density current $\mathbf{Ri} = g'h_d\cos\theta/V_d^2$ with subscript d referring to the density current is significant.

Typical velocity and density distributions of a plane density current are shown in Figure 8.10(b). The interface is defined at a location where the density gradient has the maximum. With the entrainment constant $E = V_e/V_d$ where V_e is the entrainment velocity,

$C_d = \tau_d / (\rho_d V_d^2)$ as bottom drag coefficient, and D_1 and D_2 as density profile correction coefficients, the change of *profile of the density current* may be written as

$$\frac{h_d}{3\mathbf{Ri}} \frac{d\mathbf{Ri}}{dx} = \frac{(1 + \frac{1}{2} D_1 \mathbf{Ri})E - D_2 \mathbf{Ri} \tan\theta + C_d}{1 - D_1 \mathbf{Ri}}. \tag{8.3}$$

For uniform density distribution both $D_1 = D_2 = 1$. Typically, from experiments with turbulent density currents, $D_1 = 0.6$ to 1.0, and $D_2 = 0.95$ to 1.1. Eq.(8.3) may be regarded as a generalized backwater equation for the unknown $h_d(x)$.

For equilibrium (subscript u) flow, the Richardson number remains constant and

$$\mathbf{Ri}_u = \frac{E_u + C_d}{D_2 \tan\theta - \frac{1}{2} D_1 E_u}. \tag{8.4}$$

Also, $dh_u/dx = E_u$ with the result that when accounting for the boundary condition $h_d(x = x_o) = h_o$ close to the plunging point

$$\frac{h_u - h_o}{x - x_o} = E_u. \tag{8.5}$$

The *uniform velocity* is equal to

$$V_u = \left(\frac{V_d h_d g'_d \cos\theta}{\mathbf{Ri}_u} \right)^{1/3}. \tag{8.6}$$

The term $V_d h_d g'_d$ is referred to as the *buoyancy flux*, which may remain constant. The entrainment constant is related to the Richardson number as

$$E/E_0 = \mathbf{Ri}^{-\gamma}. \tag{8.7}$$

According to Alavian, et al. (1992) the constants are $E_0 = 1.5 \times 10^{-3}$, $\gamma = 1$ for $\mathbf{Ri} > 2 \times 10^{-1}$, and $E_0 = 2.8 \times 10^{-3}$, $\gamma = 1.2$ for turbidity currents. The previous system of equations has also been established for flow in a triangular channel and on a laterally unbounded surface. An alternative approach was given by Graf (1983), and Graf and Altinakar (1993). They have also examined the nose or the front of the current. A marked similarity between density currents and dambreak waves (Chapter 10) can be noted.

8.3.4 Intrusion

If the density of a current and the stratified ambient become equal, the density current leaves the bottom and propagates horizontally into the reservoir (Figure 8.10(a)). Depending on the dimensionless number

$$\mathbf{I} = \mathbf{F}_i \mathbf{G}^{1/3} \tag{8.8}$$

where $\mathbf{F}_i$ is the densimetric Froude number at the intrusion point, and $\mathbf{G}$ the *Grashof number* involving the buoyancy frequency, the reservoir length and the average vertical eddy viscosity, three different flow regimes can be defined (Alavian, et al., 1992):

① $\mathbf{I} \geq 1$ then inertial and buoyancy forces are in equilibrium, the propagation speed is $c_i = 0.194(g_i' h_i)^{1/2}$, and the inflow thickness $h_i = 3(q_i^2/g_i')^{1/3}$;

② $\mathbf{P}^{-5/6} < \mathbf{I} < 1$ with $\mathbf{P}$ as the *Prandtl number*, then viscosity and buoyancy forces dominate; and

③ $\mathbf{I} < \mathbf{P}^{-5/6}$ then viscosity and diffusion dominate.

8.4 SEDIMENT CONTROL

8.4.1 Turbulent suspension

For a given reservoir there exists a definite relation between average sediment concentration and average stream power, defined as velocity times gradient upstream from the cross-section considered. As the sediment capacity decreases within the backwater zone of the reservoir, the sediment load decreases rapidly with distance, and material in *colloidal suspension* reaches the dam (ICOLD, 1988). Figure 8.11 shows that the vertical distribution of sediment concentration across the Verwoerd reservoir (South Africa) is nearly uniform whereas for other reservoirs a *typical density* current may be seen. Figure 8.12 shows observations on the Sautet reservoir (France).

Fine particles in reservoirs are mainly distributed due to *turbulent suspension*. The variation of concentration c in the vertical direction z under equilibrium conditions depends on the *turbulent diffusion* and may be expressed as (Rouse, 1937)

$$\frac{C}{C_a} = \left(\frac{D - z}{z} \frac{z_a}{D - z_a} \right)^m \tag{8.9}$$

where C_a is the concentration at the reference location $z = z_a$ above the bed, and the exponent m is equal to

$$m = \frac{2.1 V_s}{(gDS_e)^{1/2}}. \tag{8.10}$$

Here V_s is the settling velocity of particles considered, D the depth of flow and S_e the energy slope. With $V^* = (gDS_e)^{1/2}$ as the shear velocity, m may be regarded as a ratio of two velocities. For $m \to 0$ the

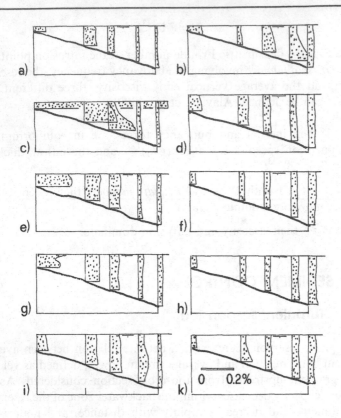

Figure 8.11
Distribution of
sediment
concentration in
Verwoerd reservoir
(SA) for various
reservoir in- and
outflows. (a) to (k)
refer to various
reservoir inflow
and outflow
configurations in
1974 (ICOLD, 1988)

variation of sediment concentration is practically negligible, and a
situation as presented in Figure 8.11(b) may result. A limit value for
the number $N = V_s/(gDS_e)^{1/2}$ is 8.3. For $N > 8.3$ the movement of
sediment through the turbulent suspension ceases. Then the sediment
concentrations along the bed are much higher than close to the surface.

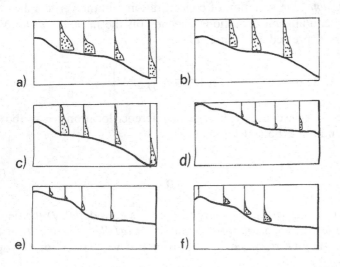

Figure 8.12
Variations of
sediment
concentrations at
Sautet reservoir
(France) in time and
space (ICOLD,
1988)

8.4.2 Turbidity currents

A *turbidity current* is developed where a layer of sediment laden fluid moves in an ambient fluid of slightly less density. The two layers are separated by a sharp discontinuity. Turbidity currents typically occur when large quantities of sediment are transported into a reservoir, such as during flood conditions where the currents are able to penetrate deeply into the reservoir.

The driving force of turbidity currents is the density difference $\Delta\rho$ associated with the sediment concentration times the bed slope S_o, and the resisting force is the density ρ times the slope of the energy line S_f. The *density number*

$$\mathbf{D} = \frac{\Delta\rho S_o}{\rho S_f} \qquad (8.11)$$

is an indicator of the relative importance of the two effects. Therefore, turbidity currents are enhanced, provided:

- density differences are large,
- flow depth is large,
- bed slope is large, and
- velocity is low.

Favourable conditions for turbidity currents are thus heavily charged floods entering a deep reservoir with a steep bed slope. Hydraulic criteria have been set up as (ICOLD, 1988)

$$\mathbf{F}_d < 1, \qquad (8.12)$$

$$\frac{S_o C^2 D^5}{Q^2} > 10^4. \qquad (8.13)$$

Eq.(8.12) involves the *densimetric Froude number* $\mathbf{F}_d = V/(g'D)^{1/2}$ where $g' = (\Delta\rho/\rho)g$ is the reduced gravitational acceleration. Accordingly, the turbidity current should remain subcritical. Eq.(8.13) gives a condition for a dimensionless number corresponding to $\mathbf{D}$ in Eq.(8.11), if a constant density ratio $\Delta\rho/\rho$ is assumed and provided the Chezy friction formula with C as friction coefficient is adopted.

The actual knowledge on turbidity currents in reservoirs is insufficient to predict sediment movements, or even the flow of sediment through a bottom outlet. For many reservoirs which were designed for a 100 year sediment storage, say, turbidity currents are not that significant because the sediment is mainly moving along the foreset slope, i.e. shifting of material from an upstream to a downstream location takes place within the reservoir.

8.4.3 Flushing of sediments

Sediments can be flushed out of a reservoir if sufficient *transport capacity* is available. In the absence of density currents, the typical concentrations during flushing are of the same order than those encountered during floods in rivers. The transport capacity depends on the grain Reynolds number $\mathbf{R}_g = (gDS_f)^{1/2}d_{50}/v$ where d_{50} is the median particle size and v the kinematic viscosity. For $\mathbf{R}_g < 13$ there is a *laminar boundary layer* in the reservoir and sediment transport occurs if (ICOLD, 1988)

$$\mathbf{N} < 0.63\mathbf{R}_g, \quad \text{for } \mathbf{R}_g < 13. \tag{8.14}$$

For a *turbulent boundary layer* that is generally developed, the condition for incipient sediment transport is

$$\mathbf{N} < 8.3, \quad \text{for } \mathbf{R}_g \geq 13. \tag{8.15}$$

These equations are valid for unconsolidated particles with $\mathbf{N} = V_s/(gDS_e)^{1/2}$ as defined previously. Velocities larger than $1\,\mathrm{ms}^{-1}$ are required to reentrain particles after deposition, whereas deposition occurs if the velocity is smaller than $0.5\,\mathrm{ms}^{-1}$. The mechanism of sediment flushing across the dam is described in 8.2.

8.5 SECONDARY HYDRAULIC EFFECTS OF RESERVOIR SEDIMENTATION

8.5.1 Upstream river

Due to the backwater of a reservoir into the approach river, the reservoir delta grows and *aggradation* of the river develops. The prediction of the delta growth is thus important in taking measures against sedimentation. The Bureau of Reclamation (Strand and Pemberton, 1982) has proposed a semi-empirical method that involves the original bottom slope and the topset bed slope. The point of intersection P between the topset and the foreset slopes is established as the median operating level of the reservoir (Figure 8.13).

Figure 8.13
Delta formation in reservoir due to sedimentation
① coarse sediments,
② original bed slope,
③ fine sediments,
④ topset slope,
⑤ pivot point P,
⑥ foreset slope,
⑦ bottomset slope,
⑧ maximum reservoir level,
⑨ normal reservoir level, ⑩ bottom outlet.

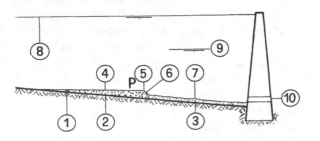

Figure 8.14
(a) Lake Ksob in Algeria, (b) lake Mead on Colorado river (USA) with the plunging front on the right side, and (c) delta of lake Havasu at Bill Williams river entry (ICOLD, 1988).

A consequence of delta formation can be the growth of grass and bushes. This anticipated dense growth of vegetation has an obstructive effect on floods and should be included in roughness estimation for the backwater prediction. Figure 8.14 shows reservoir sedimentation of lake Ksob (Algeria), lake Mead (USA) and the Bill Williams river as it enters Lake Havasu, Arizona (USA). The light colour represents the delta intrusion into the reservoir with the prominent phreatophytic growth.

8.5.2 Downstream river

The release of clear water with a capacity to pick-up a sediment load changes the stable downstream river by *degradation*. The degradation process moves progressively downstream until a new equilibrium condition is reached.

The degradation has undesirable consequences (ICOLD, 1988):

- in-channel structures such as bridges or culverts are subject to erosion,

- valuable agricultural, industrial and residential properties can be endangered within the channel banks,

- coarsening of bottom material and a change of vegetation along the channel banks with an effect on flora,

- erosion of tributary rivers due to drop of hydraulic control.

The US Bureau of Reclamation recommends two methods to counter the degradation process. The *Armouring Method* involves the application of large and coarse material that is not transported by normal river discharge. Under the armouring process the finer material is sorted out and vertical degradation slows down until the armour is sufficiently deep. The armouring depth is roughly three armouring particle diameters or 0.15 m, whichever is smaller.

The *Stable Slope Method* involves a stream slope for which the bed material is no longer transported. The resulting slope may be obtained from the Meyer-Peter and Müller (MPM) formula or Shield's diagram for no motion (Simons and Senturk, 1992; Julien, 1995). The design discharge can be the bankful flow or the 2 year flood peak discharge.

8.5.3 Disposal of sediments

Dredged sediments can be released to the river below the dam or may be discharged to the closest disposal area. Discharge to the river can result in high turbidity flows that is detrimental to ecology and recreation. Land disposal can result in spoil piles and drainage from the spoiled material can effect the surface and groundwater.

Figure 8.15
Gebidem dam
(Switzerland) with
(a) upstream end of
free surface flushing
(VAW 5213), (b)
intake cone at the
dam (VAW 3686).

Some water quality regulations prohibit the return of dredged material to the river especially out of flood periods. In the USA schemes of total containment of the water and sediment mixture have been developed (ICOLD, 1988). The concept consists of a diked pond in which the water evaporates and where the solids can be used as fill material. A common practice is to establish a vegetation to enhance the water loss through evapotranspiration. But it is sometimes difficult to find a suitable location for such a pond (topography, acceptance). The *ultima ratio* may then lead to:

- increase the dead storage of the reservoir by building a new bottom outlet and even a new intake at a higher elevation, or
- detach a bay from the reservoir by an auxiliary dam to create a disposal area.

Both solutions result of course in a loss of usable storage.

REFERENCES

Alavian, V., Jirka, G.H., Denton, R.A., Johnson, M.C., Stefan, H.G. (1992). Density currents entering lakes and reservoirs. *Journal Hydraulic Engineering* **118**(11): 1464–1489.

Breusers, H.N.C., Klaassen, G.J., Barkel, J., Van Roode, F.C. (1982). Environmental impact and control of reservoir sedimentation. *14 ICOLD Congress* Rio de Janeiro **Q54**(R23): 353–372.

Dawans, P., Charpié, J., Giezendanner, W., Rufenacht, H.P. (1982). Le dégravement de la retenue de Gebidem: Essais sur modèle et expériences sur prototype. *14 ICOLD Congress* Rio de Janeiro **Q54**(R25): 383–407 (in French).

Graf, W.H. (1983). The hydraulics of reservoir sedimentation. *Water Power & Dam Construction* **35**(4): 45–52; **35**(9): 33–38; **36**(4): 37–40.

Graf, W.H. and Altinakar, M.S. (1993). *Hydraulique fluviale*. Presses Polytechniques Universitaires Romandes, Lausanne (in French).

Groupe de Travail (1976). Problèmes de sédimentation dans les retenues. *12 ICOLD Congress* Mexico **Q47**(R30): 1177–1208 (in French).

ICOLD (1988). Sedimentation control of reservoirs. *ICOLD Bulletin* **67**. International Commission of Large Dams, Paris.

Julien, P.Y. (1995). *Erosion and sedimentation*. Cambridge University Press, Cambridge.

Rouse, H. (1937). Modern conceptions of the mechanics of turbulence. *Trans. ASCE* **102**: 463–543.

Roveri, E. (1981). Conservazione della capacità utile nei laghi artificiali. *Wasser Energie Luft* **73**(9): 199–201 (in Italian).

Simons, D.B. and Senturk, F. (1992). *Sediment transport technology*. Water Resources Publications, Fort Collins.

Strand, R.I. and Pemberton, E.L. (1982). Reservoir sedimentation. *Technical Guideline*. Bureau of Reclamation, Denver.

a)   b)

Earth slides Morignone, Valtellina Italy on July 28, 1987. Slide volume 40 Mio m^3, slide width 800m, height difference 1200m

9

Impulse Waves from Shore Instabilities

9.1 INTRODUCTION

Impulse waves can be generated by various effects, such as rock falls, land slides, ice falls, glacier calvings or snow avalanches. Such waves may endanger a dam and the reservoir due to wave runup, or even overtopping of the dam crest and the resulting floods in the downstream valley. Also, impulse waves can lead to significant erosion of a shore which can result in secondary events (Vischer, et al., 1991).

There exists a significant body of knowledge on the mechanics of wave generation, wave propagation along the reservoir and wave runup or overtopping. The results refer mainly to plane flow and these indications can be accounted for in spatial situations. It is clearly understood that the phenomenon is governed by the Froude similarity law and the hydraulic model is currently an approach that should be considered for all projects where impulse waves are a concern.

The following discussion is mainly a summary of studies conducted at VAW, ETH Zurich, and refers to plane impulse waves. Also, an account for 3D impulse waves is given where available. All three stages of flow, including wave generation, wave propagation, and wave runup are accounted for.

9.2 IMPULSE WAVES

9.2.1 Wave theories

Waves are amenable to mathematical treatment, and there exists a large body of literature relating to water waves (Wiegel, 1964). The main parameters of water waves are (Figure 9.1(a)):

- *wave amplitude* a_M defined as the crest elevation over the undisturbed water line,
- *wave length* L_w between two adjacent wave crests, and

- *water depth* h_o as the elevation of the still water level over the bottom.

Depending on the *wave shallowness* $\tau = a_M/h_o$ and the *wave steepness* $\epsilon = (h_o/L_w)$ various wave types may be distinguished (Figure 9.1(b)):

① *linear waves* where both $\tau \ll 1$ and $\epsilon^2 \ll 1$,

② *Airy waves* where $\tau \gg \epsilon^2$ and the wave is sinusoidal,

③ *Boussinesq waves* with $\tau \approx \epsilon^2$ where curvature effects become significant and the waves are non-linear.

④ *dispersive waves* where $\tau \ll \epsilon^2$.

Usually, these wave theories account only for the gravitational forces and the effects of viscosity and surface tension are neglected. Impulse waves as treated in this chapter are typical phenomena where gravity is the dominant parameter. They are normally intermediate between the deep water waves and the shallow water waves with $\tau \approx \epsilon^2$. The *solitary wave* is a particular wave type of relevance to impulse waves. A general account of water waves is given, e.g. by Le Mehauté (1976).

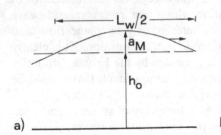 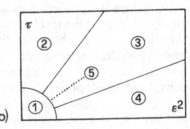

a)   b)

Figure 9.1
Wave theories (a) schematic wave propagation, (b) Effect of wave shallowness $\tau = a_M/h_o$ and steepness $\epsilon = h_o/L_w$ on wave theories (Sander and Hutter, 1992)

9.2.2 Impulse waves

Waves generated in a reservoir include three mechanisms (Figure 9.2):

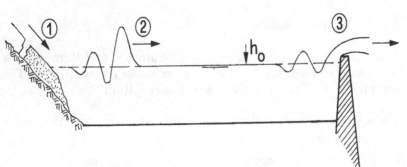

Figure 9.2
Impact wave mechanisms on reservoir, schematic

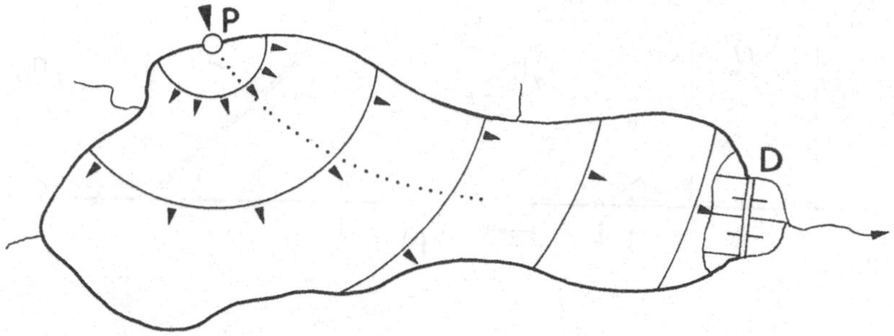

Figure 9.3
Impulse wave in reservoir generated at point P and propagating towards a dam D

① wave generation by movement of a mass,

② wave propagation in the reservoir, and

③ wave impact on topography, such as opposite shore or dam, with possible wave overtopping.

The waves may be generated by various sources, including rock, ice or snow avalanches, glacier calving, shore instabilities, and earthquakes, among others. The characteristics of *plane* impulse waves may be summarized as follows:

- The first wave height corresponds normally to the maximum elevation (Noda, 1970, Wiegel, et al., 1970). Also the first wave contains more energy than all the following waves.

- The waves decrease in height as they propagate over a nearly horizontal bottom (Hunt, 1988).

- For each wave that has travelled over several of its wavelengths, the wave pattern depends only on the bathymetry of the reservoir and its volume.

- Dispersive waves generated by a horizontal mass displacement transform with increasing propagation length to a solitary wave, followed by a sinusoidal wave train (Miller, 1960).

Spatial impulse waves are more complex and the results may not be generalized so easily. The main parameters are the Froude number based on the impact velocity, the volume of impact, the impact angle with respect to the reservoir axis and the distance from the impact location. Figure 9.3 shows a typical event with the impact location P, and the wave generation in the reservoir.

9.3 GENERATION OF IMPULSE WAVES

9.3.1 Wave generation by moving wedge

The momentum transfer of masses falling into a reservoir to the generation of impulse waves has not yet been systematically analysed. The

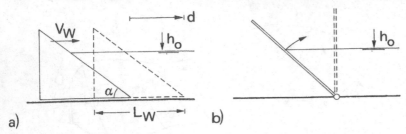

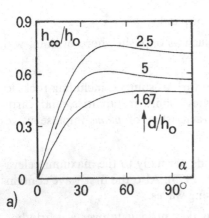

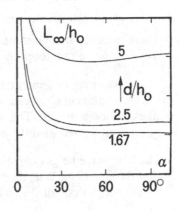

Figure 9.4
Experimental
arrangement of
Sander and Hutter
(1992) (a) moving
wedge, (b) rotating
plate

Figure 9.5
Scaling parameters
for waves
generated by
moving wedge (a)
maximum possible
wave amplitude
ratio h_∞/h_o and (b)
corresponding
wave length ratio
L_∞/h_o as functions
of wedge angle α
and relative
dislocation d/h_o
(Sander and Hutter,
1992)

effects of mass fracturing and air entrainment complicate the
approach.

However, Sander and Hutter (1992) have considered a somewhat
similar problem: What are the wave characteristics for a moving body
submerged in the reservoir? Moving wedges (subscript W) having a
translation velocity V_W or rotating plates were used. The determining
parameters of the waves so generated are the relative dislocation d/h_o,
the wedge Froude number $\mathbf{F}_W = V_W/(gh_o)^{1/2}$ and the wave period
$T_w = t_w(g/h_o)^{1/2}$ where h_o is the still water depth (Figure 9.4).

For $0.1 < \mathbf{F}_W < 1.1$ the waves generated have always the same
characteristics: a steep front extending to a first wave maximum is
followed by a slightly slower decay in the first wave tail. Usually, a
wave train was produced that began with a *compact front wave* and a
wave tail ordered according to the wave amplitudes.

The maximum possible wave height h_∞ relative to the initial flow
depth h_o, and the corresponding wave length L_∞/h_o depend on the
wedge angle α displaced by d/h_o over the bottom of the rectangular
channel. Both h_∞ and L_∞ serve as scaling values and are plotted in
Figure 9.5.

For a given wedge Froude number $\mathbf{F}_W$, wedge angle α, relative
dislocation d/h_o, and thus known scalings h_∞ and L_∞, the maximum
wave amplitude a_M/h_∞ and the corresponding wave length ratio

Figure 9.6
(a) Maximum wave
amplitude a_M/h_∞ as
a function of
$F_W(h_o/h_\infty)$, (b)
corresponding
wave length ratio
L_M/L_∞ as a function
of $F_W(h_o/L_\infty)$
(Sander and Hutter,
1992)

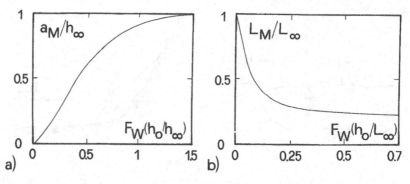

L_M/L_∞ may be obtained from Figure 9.6. These plots clearly indicate that the governing parameter of wave propagation is $F_W(h_o/h_\infty)$, and $F_W(h_o/L_\infty)$. For a given wedge angle, the wave amplitude ratio a_M/h_∞ increases with increasing wedge velocity and decreases as h_∞ increases. To the lowest order of approximation, one may even write the linear relation $a_M/h_\infty = F_W(h_o/h_\infty)$, provided that $a_M/h_\infty < 1$. Thus, independent of the wedge angle α, the result is simply

$$a_M/h_o = F_W. \qquad \bullet \quad (9.1)$$

Accordingly, the wave amplitude increases proportionally with F_W and h_o.

9.3.2 Wave generation by falling mass

According to Huber (1982), Switzerland has been endangered almost during every decade of this century by impulse waves occurring in natural lakes. The damage of property has been substantial, and people have lost their lives in two cases. Other occurrences in reservoirs, especially in Italy and China, have resulted in large catastrophes.

Rock avalanches or bank slides in steep reservoir shores include various problems, such as the form and extent of the slide, the mass transfer on to the water and the main characteristics of the impulse waves generated (Vischer, 1986). Accordingly, these topics have to be addressed by experts in geology and wave mechanics. The latter part is dealt with hereafter, based on the knowledge of the main parameters of the falling mass. These include the slide volume V_s of density ρ_s, the still water depth h_o, the slope angle γ at the impact site, and the distance x from the impact location (Figure 9.7). All points where wave heights have to be predicted should be visible from the impact site. Indirect waves are influenced by wave reflection and diffraction and are much smaller than direct waves. In the following, only direct waves are considered (Figure 9.7).

The *range of validity* of the test results is (Huber, 1982):

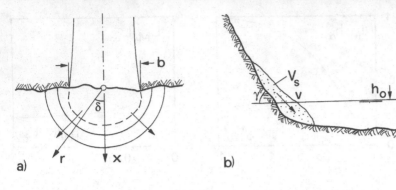

Figure 9.7
Definition plot for wave generation in reservoirs (a) plan, (b) section

- relative wave height smaller than wave breaking limit, i.e. $a_M/h_o < 0.78$;

- relative propagation domains $5 < x/h_o < 100$ for 2D waves, and $5 < x/h_o < 30$ for 3D waves;

- slide velocity larger than about 50% of the wave celerity c;

- slide angle γ between 28° and 60°. For $\gamma < 25°$, friction normally inhibits the sliding of material;

- slide mass is a dense debris flow. From a compact rock waves would be higher, and a completely disintegrated mass resulting from a blasting yields much smaller waves.

The data of Huber have been reanalysed and can be described within ±15% (Huber and Hager, 1997)

$$\frac{a_M}{h_o} = 0.88 \sin\gamma \left(\frac{\rho_s}{\rho_w}\right)^{1/4} \left(\frac{V_s}{bh_o^2}\right)^{1/2} \left(\frac{h_o}{x}\right)^{1/4}. \qquad (9.2)$$

The following statements apply:

1. The slide angle γ at the impact site has a major effect on the relative wave height, followed by the relative volume $V_s/(bh_o^2)$ where b is the width of the slide. The effects of density and location are relatively small.

2. The effect of *slide Froude number* normalized with the still water depth h_o is insignificant. Accordingly, a rapid slide in a deep reservoir produces nearly the same wave height as a slower slide in a correspondingly shallower reservoir.

3. The *wave propagation velocity* can be approximated with the expression for surge or solitary waves as

$$c^2 = g(h_o + a_M). \qquad (9.3)$$

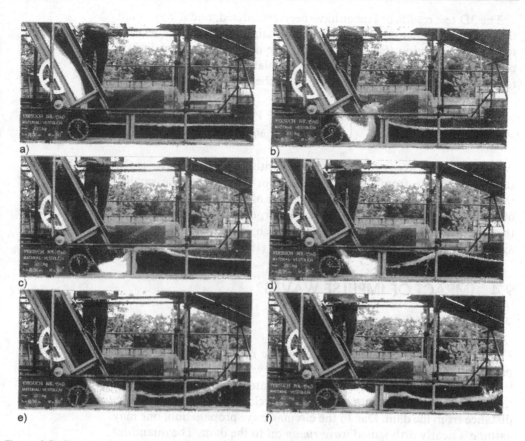

Figure 9.8 Temporal development of impulse wave due to slide (Huber, 1982)

The problem of large impulse waves may thus be approximated with the Boussinesq equations, given that the effects of streamline curvature are significant (Figure 9.8).

9.3.3 Spatial impulse waves

For *spatial* impulse waves, the following comments can be added:

- The wave propagation is semi-radial, from the impact location towards the reservoir (Figure 9.3).

- The highest wave location is in the extension of the slide direction, and lateral waves at equal distance from the impact site are considerably lower.

- The decay of wave maximum corresponds to the plane experiments. Huber (1982) gives numbers in a table for the estimation of the wave decay.

The 3D test results of Huber have been reanalysed, and it was found that the wave reduction in a pool depends mainly on the angle δ relative to the slide direction (Figure 9.7). Also, the decay of wave height in a reservoir is significantly larger than in the plane flume. The test results may be expressed to $\pm 20\%$ accuracy as (Huber and Hager, 1997)

$$\frac{a_M}{h_o} = 2 \cdot 0.88 \sin\gamma \cos^2\left(\frac{2\delta}{3}\right) \left(\frac{\rho_s}{\rho_w}\right)^{\frac{1}{4}} \left(\frac{V_s}{bh_o^2}\right)^{\frac{1}{2}} \cdot \left(\frac{r}{h_o}\right)^{-\frac{2}{3}} . \qquad (9.4)$$

Accordingly, the decay of wave height is with $(h_o/r)^{2/3}$, where r is the radial coordinate from the impact site. For lateral angles δ smaller than $\pm 20°$ there is practically no reduction of wave height. The effect of variable reservoir depth, h_o, can be incorporated by accounting for the average depth. An alternative approach, and computational examples are presented by Huber and Hager (1997).

9.4 IMPACT OF IMPULSE WAVES

9.4.1 Governing parameters

Figure 9.9 shows a *plane* impulse wave on a nearly horizontal bottom propagating towards a dam. The maximum wave height from the wave trough to the wave peak is h_M, the still water depth is h_o, and L_w is the wave length. The front propagation velocity is c_w, and x is the distance from the dam. Due to the circular wave propagation, one may assume a nearly orthogonal *wave runup* on to the dam. The quantities to be determined are the runup height r, the overtopping volume V_d if the runup height r is larger than the freeboard height f, and the time of overtopping t_d.

Based on a dimensional analysis, Müller (1995) showed that both the relative runup height $R = r/h_o$ and the overtopping volume V_d/h_o^3 depend on the relative wave height h_M/h_o, the wave steepness $\epsilon = a_M/L_w$, the wave period $\tau = T_w/(h_o/g)^{1/2}$ with $T_w = L_w/c_w$ and c_w as wave propagation velocity, and the runup angle β. Also, for sufficiently large waves, the Froude similarity law is governing impact waves.

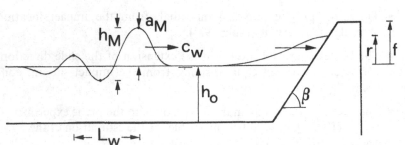

Figure 9.9
Wave propagation and runup on dam, definition of variables

9.4.2 Runup mechanism

Runup waves are well-known either as *wind waves* with a short wave length, or as *tsunamis* for long wave lengths. Whereas wind waves are periodical in deep water, tsunamis are long waves consisting of a solitary front wave followed by a periodic wave train. Herbich (1990) distinguished between the non-breaking and the breaking tsunamis, where the latter may break far away or close to the coast. These waves occur typically in the Pacific ocean and are notorious for their destructive action.

The impulse wave considered belongs to waves in the transition regime with a wave shallowness h_o/L_w between 0.5 and 2, with properties close to shallow water flows. Accordingly, the wave parameters are variable with location and time. On reaching the shore, the wave piles up and may possibly break. Impulse waves are thus comparable to tsunamis. Figure 9.10 refers to runup on a dam with an upstream slope 1:3. Breaking is not seen to occur.

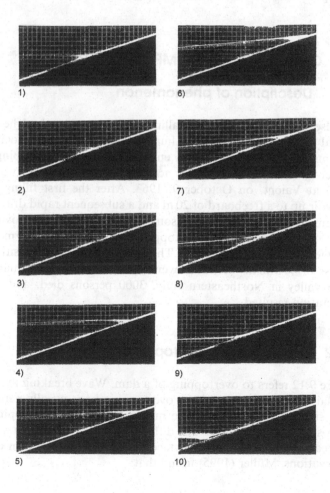

Figure 9.10
Runup on smooth dam (Müller, 1995). Note that wave breaking does not occur

9.4.3   Runup height

The relative runup height $R = r/h_o$ was previously related to the relative wave height h_M/h_o, the wave steepness $\epsilon_A = a_M/L_w$, the wave speed $\tau = T_w(g/h_o)^{1/2}$, and the runup angle β (Figure 9.9). From roughly 200 experimental runs the effect of wave period was found to be nearly negligible and

$$R = 1.25(\pi/2\beta)^{0.2}(h_M/h_o)^{1.25}(h_M/L_w)^{-0.15}. \qquad (9.5)$$

Increasing the runup angle β decreases the runup height R. In the domain of experiments $18° \leq \beta < 90°$, this effect gives maximum variations up to 40%. The relative wave height is the governing parameter, and the experiments refer to the domain $0.01 \leq h_M/h_o \leq 0.51$. The wave steepness has a relatively small effect in the experimental domain $0.001 \leq h_M/L_w \leq 0.0135$. Waves with $B_w < 3$ are breaking as they run up a shore, where $B_w = \tan\beta/(h_M/L_w)^{1/2}$ is the *wave breaking index*.

9.5   OVERTOPPING OF IMPULSE WAVES

9.5.1   Description of phenomenon

Impulse waves generated by falling masses can reach a height of several meters, wave lengths of hundreds of meters and velocities up to 30 m/s. Accordingly, they may endanger dams by overtopping or by impact action. The most tragic accident occurring in this way happened at Vaiont, on October 9, 1963. After the first filling of the reservoir up to a freeboard of 20 m and a subsequent rapid drawdown, an immense rock avalanche of some 300 Mio m^3 occurred over 2 km width. The arch dam was overtopped with a height of 100 m, which damaged the crest (Figure 9.11). The dam remained at its position, but a mass of 40 Mio m^3 of water overtopped the dam and damaged the Piave valley in Northeastern Italy. 3000 persons died, and terrible devastation resulted.

9.5.2   Description of overtopping

Figure 9.12 refers to overtopping of a dam. Wave breaking in front of the dam is infrequent, and the overtopping jet is usually aerated on both sides. Approximately, one may consider the overtopping as a quasi-steady spillway flow. The differences are mainly in the modified upstream flow with a rising surface towards the dam. From detailed observations, Müller (1995) found that:

Figure 9.11
Arch dam of Vaiont
(Italy) after
overtopping of a
100 m high impulse
wave

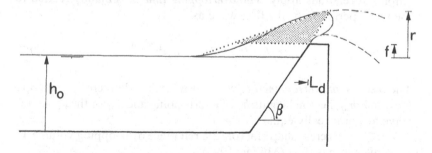

Figure 9.12
Overtopping of
dam with (···)
overtopping
volume (Müller,
1995)

- the wave front is less steep than the wave tail, and
- the maximum overtopping level occurs after (2/3) of the overtopping period.

A generalized wave profile was described.

9.5.3   Characteristics of overtopping

The mechanism of overtopping is so complex that a simplified approach for the volume of reference must be given. Müller (1995) has related the overtopping volume V_d to the reference volume V_o which would occur for zero freeboard. Clearly, the relative overtopping volume V_d/V_o decreases from the maximum value for zero freeboard $f = 0$ to the runup height r. Müller's result for the envelope of experimental data

$$V_d/V_o = (1 - f/r)^2 \qquad (9.6)$$

can easily be explained when assuming that the shape of the free surface just upstream from the dam is nearly a triangular wedge of height $r - f$, and the corresponding maximum value has a height equal to the runup height r (Figure 9.12). Figure 9.13 shows a series of photographs that is analogous to Figure 9.10 for a broad-crested dam of impact slope 1:1.

The *maximum overtopping volume* V_o for waves on to a dam without freeboard, i.e. when the reservoir is completely filled to the crest, depends on the:

- angle β of the dam cross-section on the water side,
- crest geometry (i.e. sharp-, round-, or broad-crested),
- overtopping time t_o,
- maximum overtopping height h_d, and
- shallowness parameter of approach flow.

According to the experiments of Müller (1995), the following semi-empirical relations apply. The *overtopping time* t_o is mainly related to the wave period t_w of the first wave as

$$t_o(g/h_o)^{1/2} = 4[t_w(g/h_o)^{1/2}]^{4/9}. \qquad (9.7)$$

The *maximum overtopping height* h_d at a vertical dam crest is equal to $h_d = 0.96 h_M$, i.e. the heights h_d of overtopping and h_M of the approach wave are practically equal.

With c_o as a crest shape factor, the reference overtopping volume V_o per unit width is thus (Müller, 1995)

$$V_o = \sqrt{2} c_o (g h_M^6 h_o^2 t_w^2)^{2/9}. \qquad (9.8)$$

The parameter V_o depends strongly on the maximum wave height h_M of the approach flow, whereas the effects of still water depth h_o, and

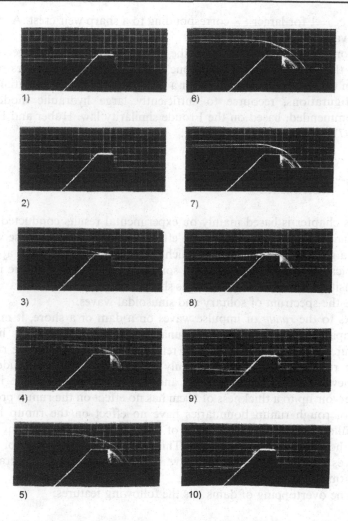

Figure 9.13
Overtopping of
broad crested dam
(Müller, 1995)

wave period t_w are modest. The accuracy of Eq.(9.8) is ±10%. The *crest shape coefficient* $c_o = c_\beta \cdot c_\xi$ in Eq.(9.8) depends on the dam angle β, the crest width ratio $\xi_d = h_d/L_d$ and on the base value $c_{90°}$ for the sharp-crested standard weir (Figure 9.12). The effect of dam angle β may be approximated as

$$c_\beta = c_{90°} + 0.05\sin\left[\frac{2}{3}(90° - \beta)\right]. \tag{9.9}$$

A maximum occurs for $\beta \approx 30°$, as was found from various experiments (Hager, 1994).

The effect of relative *crest width* $\xi_d = h_d/L_d$ on the overtopping volume for $0.2 < \xi_d < 2.1$ may be expressed as

$$c_\xi = 0.60 + 0.19\xi_d \tag{9.10}$$

and $c_\xi = 1$ for larger ξ_d, corresponding to a sharp weir crest. A standard value is $c_\xi = 1$.

For known characteristics of the approach wave, the dam geometry and the reservoir topography, one may thus predict estimates of the main overtopping parameters for a nearly plane flow. For spatial flow configurations, recourse to sufficiently large hydraulic models is recommended, based on the Froude similarity law. Huber and Hager (1997) present a typical example.

9.6 REMARKS

This chapter is based mainly on experimental results conducted on a sufficiently large scale that scale effects are negligible. Impulse waves are a gravity phenomenon in which the Froude similarity law applies. Typical wave periods are 1/2 to some minutes. The waves are in the transition wave zone or behave as shallow water waves. The wave type is in the spectrum of solitary and sinusoidal waves.

As to the *runup* of impulse waves on a dam or a shore, it may be compared with the runup of tsunamis. Wave breaking does hardly occur, because the wave is fully reflected, or it breaks before runup. The runup height depends mainly on the approach amplitude and somewhat on the runup angle and the wave steepness. Ice in the reservoir up to a thickness of 50cm has no effect on the runup process. Also, rough runup boundaries have no effect on the runup height (Müller, 1995). The bathymetry of a reservoir has a significant effect on the runup of impulse waves. This chapter applies mainly for plane waves, since spatial waves have not yet received a generalized-treatment.

The overtopping of dams has the following features:

- the *maximum height* of the overtopping wave over a dam crest is always smaller than the maximum height of the corresponding runup wave,

- the *overtopping volume* related to the reference volume at zero freeboard increases quadratically with decreasing freeboard,

- the *impulse wave profile* normalized by the wave amplitude and the wave length is characteristic both during the wave propagation in the reservoir and during the overtopping,

- the *reference overtopping volume* V_o depends mainly on the approach wave amplitude and partly on the reservoir depth and the wave period, and

- *spatial features* of overtopping waves are complex. Hydraulic modelling is then strongly recommended.

REFERENCES

Hager, W.H. (1994). Dammüberfälle (Dam overfalls). *Wasser und Boden* **46**(2): 33–36 (in German).

Herbich, J.B. (1990). *Handbook of coastal and ocean engineering* **1**. Gulf Publishing Company, Houston.

Huber, A. (1982). Impulse waves in Swiss lakes as a result of rock avalanches and bank slides. *14 ICOLD Congress*, Rio de Janeiro **Q54**(R29): 455–476.

Huber, A., and Hager, W.H. (1997). Forecasting impulse waves in reservoirs. *18 ICOLD Congress*, Florence, **C31**: 993–1005.

Hunt, B. (1988). Water waves generated by distant landslides. *Journal Hydraulic Research* **26**(3): 307–322.

Le Mehauté, B. (1976). *An introduction to hydrodynamics and water waves.* Springer, New York.

Miller, D.J. (1960). Giant waves in Lituya Bay. Geological Survey, *Professional Paper* **354**(C): 51–83.

Müller, D.R. (1995). Auflaufen und Überschwappen von Impulswellen an Talsperren (Runup and overtopping of impulse waves at dams). *Dissertation* **11113**. Eidg. Technische Hochschule ETH, Zürich (in German).

Noda, E. (1970). Water waves generated by landslides. *Journal Waterways, Harbors and Coastal Engineering Division* ASCE **96**(WW4): 835–855.

Sander, J. and Hutter, K. (1992). Evolution of weakly non-linear shallow water waves generated by a moving boundary. *Acta Mechanica* **91**: 119–155.

Vischer, D.L. (1986). Rockfall-induced waves in reservoirs. *Water Power and Dam Construction* **38**(9): 45–48.

Vischer, D., Funk, M., Müller, D. (1991). Interaction between a reservoir and a partially flooded glacier. Problems during the design stage. *18 ICOLD Congress* Vienna **Q64**(R8): 113–135.

Wiegel, R.L. (1964). *Oceanographical Engineering.* Prentice-Hall, London.

Wiegel, R.L., Noda, E.K., Kuba, E.M., Gee, D.M., Tornberg, G.F. (1970). Water waves generated by landslides in reservoirs. *Journal Waterways, Harbors and Coastal Engineering Division* ASCE **96**(WW2): 307–333; **97**(WW2): 417–423; **98**(WW1): 72–74.

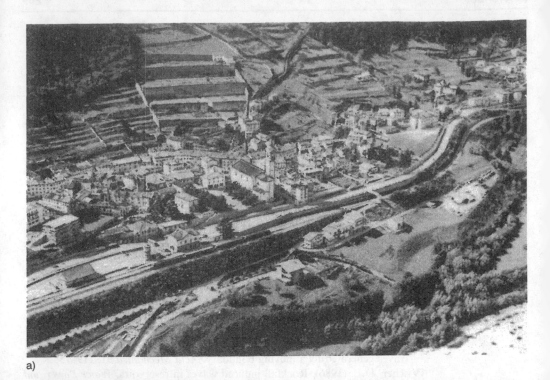

a)

b)

Vaiont dam disaster, tailwater village Longarone (a) prior and (b) after dam overtopping on October 9, 1963

10

Dambreak Waves

10.1 INTRODUCTION

A dam failure may release large quantities of water that can create major flood waves in the tailwater and cause serious damages. Singh (1996) has noted 1000 dam failures since the 12th century, and about 200 have occurred in the 20th century so far causing a loss of more than 8000 lives and damage worth millions of dollars. The annual probability of dam failure is estimated at 10^{-4} and the probability of failure during the lifetime of a dam (100 years) is 10^{-2}.

Schnitter (1993) demonstrated by a data collection of ICOLD, that the percentage of embankment dams failing in a certain year has dropped at least tenfold during the first half of this century. The dams thus exhibit remarkable longevity, and have currently a good safety record, compared to other types of structures. *Dam safety* is now a significant issue in dam engineering, and the study of consequences of dam breaking has been demanded by authorities for various existing dams.

A dam can be threatened by factors such as:

- floods,

- rockslides or landslides,

- earthquakes,

- deterioration of heterogeneous foundation,

- poor quality of construction and construction materials,

- differential settlement,

- improper reservoir management, and

- acts of war.

Singh presents an exhaustive list of causes for dam failure.

The *causes of dam failure* are mainly (Singh, 1996):

- 30% (±5%) floods exceeding spillway capacity,

- 37% (±8%) foundation problems (seepage, piping, excess pore pressures, fault movement, settlement),

- 10% ($\pm$5%) slides (earth, rock, glacier, avalanches), and

- 23% ($\pm$12%) improper design and construction, inferior quality of material, acts of war, lack of operation and maintenance.

Foundation problems and overtopping are thus the most important reasons for dam failures. Also, the probability of failure is much greater for *embankment dams* than for concrete and masonry dams. After 1900, almost half of the failures were due to overtopping (Schnitter, 1993). In 41% of those cases, the spillway was underdesigned whereas in 21% overtopping was caused by problems with spillway gate operation.

10.2 DAM BREACHING

10.2.1 Breaching characteristics

Dams can fail either gradually of instantaneously. The type of failure depends on the cause of failure and the dam type. When a dam fails instantaneously, i.e. a large portion or the entire dam is removed within a short time, a sudden release of water generates a flood wave propagating in the tailwater valley. Because the baseflow in such a case can often be neglected, the dambreak wave propagates practically over a dry bed. A negative wave is created upstream propagating up along the reservoir. The topography of the reservoir controls the negative wave. Since an instantaneous failure constitutes the most adverse condition, it is commonly adopted for dambreak modelling and prescribes the upper bounds for the expected damages. Concrete dams that fail by overtopping or sliding, are typical examples of *instantaneous failure*. This failure mechanism is dealt with in Section 10.5.

Earth dam disasters are often due to *gradual failure* over a period of time. The duration of failure is typically several hours. For earthdams, the assumption of instantaneous failure is unrealistic. Then no shock front is developed at the wave tip and the flow is gradually varied. This failure mechanism is dealt with in this section.

10.2.2 Examples of dam breachings

An example of *instantaneous dam failure* occurred on December 2, 1959 at *Malpasset dam* in Southern France (James, 1988). The collapse occurred in a burst, that destroyed practically the entire dam, resulting in the loss of several hundreds of lives and millions of dollars in property damage. According to experts, the failure was unpredictable.

Malpasset dam was located in a narrow gorge of the Reyran river 12 km upstream of Fréjus, a resort community on the French Riviera.

The dam was a double curvature arch dam, 66 m high with a crest length of 222 m. The thickness varied from 6.8 m at the base to 1.5 m at the crest. The dam had an ungated spillway 30 m long and a bottom outlet 1.5 m in diameter controlled by a butterfly valve, and a reinforced concrete apron to protect against downstream scour.

Shears and faults are numerous in the dam foundation, and are orientated in the same directions as the jointing. The foundation was not grouted, except in the contact zone immediately below the concrete blocks. East of the river, the dam rested on a *wedge of gneiss* and it was not known that this wedge was detached from the underlying rock mass by two converging discontinuities. The geological investigations included the inspection of the site by a professor of geology, eight grout holes 30 m in depth, and examination of the foundation excavation. This investigation was based on the supposition that a gravity dam would be constructed. It is not clear whether the geologist was duly informed when the design was changed to an arch dam.

The construction of the dam started in 1952 and reservoir filling began in late 1954. Exceptional floods in November 1959 filled the last 4 m of storage within three days. Some cracks were observed in the concrete apron a few weeks before dam failure. In the afternoon of the day of failure, a group of engineers decided to open the bottom outlet to control the rising reservoir pool. At 6 p.m. the caretaker left the crest of the dam from a routine job. At 9.13 p.m. Electricité de France registered a power outage of the Malpasset 10 kV line. The first failure of a modern concrete arch dam had occurred catastrophically. The witness closest to the scene was about 1.6 km downstream and reported first feeling a trembling of the ground, followed by a loud, brief rumble, then a strong blast of air. Finally, the water arrived in two pulses, a wave that overflowed the stream banks and then a wall of water, which the witness barely escaped.

Figure 10.1(a) shows remnants of the dam. According to *Coyne et Bellier*, the designers of the dam, the failure was due to secondary alteration of the rock at the left abutment. As the reservoir filled, the hydrostatic force increased against the "underground dam" that had been created by compression of the rock abutment. Ultimately, a crack opened in the foundation along the upstream face of this barrier and the base of the rock wedge was exposed to full reservoir pressure. Unable to withstand this added force, the rock wedge slid outward and upward along the plane of the fault. The vertical component of this force raised the dam, allowing it to be rotated intact as though hinged at its extremity at the right abutment (Figure 10.1a). Experts agreed that conventional grouting and drainage probably would not have provided protection, considering the geologic structure and rock textures at the site.

An example for a *gradual dam failure* occurred to *Teton dam* (Idaho). The dam is located in a steep walled canyon at Rexburg Bench (Jansen, 1988). In the reservoir area consisting of volcanic

a)

b)

Figure 10.1
Remnants of
(a) Malpasset dam
(France) and
(b) Teton dam (USA)

rock, the permeability is high and conveys away an appreciable percentage of the water stored, with significant recharge of the regional groundwater table. The dam has a compacted, central core, zoned earth and gravel fill embankment. Its height above bedrock was 126 m. The dam was 950 m long at the crest, and had nearly 3H to 1V slopes, with a total fill volume of $7.65 \times 10^6 m^3$.

Dam construction was from April 1972 to November 1975. Storage started in December 1975, with the auxiliary low level outlet (tunnel under the right abutment) in operation. The releasing discharge was $8.5 m^3 s^{-1}$ until May 1976. Due to heavy spring runoff, the discharge was increased finally to $27\text{m}^3\text{s}^{-1}$, somewhat in excess of its rated capacity. The combined capacity of the outlet and the main river outlet works located under the left abutment was $120\text{m}^3\text{s}^{-1}$. The river outlet was incomplete, however, up to the time of dam failure, and hence was unavailable for participation in planned control of the reservoir filling schedule.

Teton dam (Idaho) failed during the first filling at a reservoir depth of 84 m, 7 m below of maximum normal reservoir level. Between 7 a.m. on June 5, 1976, when the initial damaging leaks were first seen in the right groin of the dam, and noon of that day, total breaching and failure occurred, beginning with the appearance of muddy springs, followed quickly by piping through the embankment, and ending with collapse of the crest into the rapidly enlarged "pipe". By 6 p.m. the reservoir was virtually empty, having released $3 \times 10^6 \text{m}^3$ water with a peak outflow of $28\,000\text{m}^3\text{s}^{-1}$. It was concluded that the cause of failure was the inadequate protection of parts of the impervious core material from internal erosion (Jansen, 1988).

A number of dam failures including details of causes and damages are also summarized by Singh (1996).

10.2.3   Breach characteristics

Based on 52 historical dam failures, Singh (1996) has summarized the main characteristics of dam breaches. The *breach shape* can be approximated in all cases as trapezoidal, with a ratio between the top and the bottom width of 1.29($\pm$0.18), and extreme values from 1.06 to 1.74. The ratio of top width B to depth of breach d depends linearly on the ratio of H_s/H_d, where $H_s = V_s^{1/3}$ with V_s as the storage volume (Figure 10.2). For the dams mentioned by Singh one can correlate $B/d = 0.40 H_s/H_d$, provided $H_d > 8m$ (Figure 10.2(a)). The angle between the breach side slope and the vertical was in the majority of cases between 40° and 50°. The *failure time* was between half an hour and 12 hours. For most of the cases the failure time was less than three hours. With a probability of 50%, the failure time was less than 90 min, for the data considered by Singh (1996). For $1 < H_s/B_a < 10$, the time of failure t_f is nearly $(g/H_d)^{1/2}t_f = 1.5H_s/B_a$ where B_a is the average

Figure 10.2
Empirical correlations for breach characteristics (a) relative top width and (b) relative peak discharge as functions of relative dam height

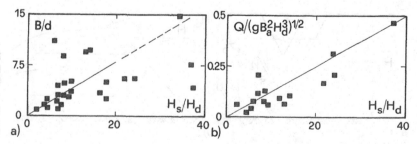

failure width. The relative *peak discharge* Q_p can be related to the relative dam height as $Q_p/(gB_a^2H_d^3)^{1/2} = 1.25 \times 10^{-2}(H_s/H_d)$, for dams that failed after 1925 (Figure 10.2(b)). These relations can be used for preliminary analysis of the dam breach characteristics, i.e. the breach geometry and the outflow hydrograph. It is noted that both the typical storage depth H_s and the dam height H_d are the significant parameters in the analysis.

10.2.4 Singh's model for dam breach development

A dam breach can be modeled as a *two-phase water-sediment flow* by using the water-volume balance

$$A_s(H)\frac{\mathrm{d}H}{\mathrm{d}t} = Q_b \tag{10.1}$$

plus the erosion rate $\mathrm{d}z/\mathrm{d}t$ as a function of velocity across the breach

$$\frac{\mathrm{d}z}{\mathrm{d}t} = -\alpha u^\beta. \tag{10.2}$$

Here, A_s is the reservoir surface, H the reservoir surface elevation from a reference datum, t is time and Q_b the breach discharge (Figure 10.3). Further $z = z(t)$ is the breach bottom, u the average breach velocity and $\alpha[ms^{-1}]^{1-\beta}, \beta[-]$ coefficients. Eq.(10.1) holds if the difference between the reservoir inflow and the outflows over the spillway, the bottom outlet and the powerhouse are much smaller than the breach discharge. During the event, essentially all discharge flows across the breach section, therefore.

The breach discharge is equal to the breach velocity times the breach section, i.e. $Q_b = uA_b$, with $u = [2gC_d^2(H - z)]^{1/2}$ where C_d is the discharge coefficient. Eliminating the time differential gives a differential equation for $H(z)$

$$\frac{\mathrm{d}H}{\mathrm{d}z} = \alpha^{-1}(A_b/A_s)[2gC_d^2(H - z)]^{(1-\beta)/2}. \tag{10.3}$$

The breach shape can be approximated as trapezoidal with a bottom width b and a side slope 1(vertical) : s(horizontal), i.e.

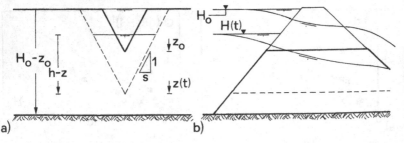

a)   b)

Figure 10.3
Definition of dam breaching for embankment dams (a) front view, (b) section

$$A_b = b(H - z) + s(H - z)^2. \tag{10.4}$$

For $s = 0$ the breach shape is rectangular, and for $b = 0$ it would be triangular (Figure 10.3). To reduce the number of parameters, and to simplify the approach, only the latter case is considered here ($b = 0$). Inserting in Eq.(10.3) and using the transformation $h = H - z$ gives

$$1 + \frac{dh}{dz} = \frac{s}{\alpha A_s}(2gC_d^2)^{(1-\beta)/2}h^{(5-\beta)/2}. \tag{10.5}$$

This must be solved subject to the two initial conditions $H = H_o$ and $z = z_o$ at $t = 0$. With the dimensionless parameters $H = h/h_o$ and $Z = z/z_o$, and the constants $\gamma = h_o/z_o > 1$ and $C^2 = \alpha A_s$ $(2gC_d^2 h_o)^{(\beta-1)/2}/(sh_o^2)$, Eq.(10.5) reads

$$1 + \gamma\frac{dH}{dZ} = C^{-2}H^{(5-\beta)/2}. \tag{10.6}$$

For $H(Z = 1) = 1$, the solutions read for the linear ($\beta = 1$) and the quadratic ($\beta = 2$) erosion rate laws ($\delta = C^{2/3}$)

$$\beta = 1 \quad Z(H) = 1 - \frac{1}{2}\gamma C^2\ln\left[\frac{C + H}{C - H}\cdot\frac{C - 1}{C + 1}\right], \tag{10.7}$$

$$\beta = 2 \quad Z(H) = 1 - 2\sqrt{3}\gamma\arctan\left[\frac{6\delta(H^{1/2} - 1)}{3\delta^2 + (2H^{1/2} + \delta)(2 + \delta)}\right] +$$

$$\frac{\gamma}{3\delta}\ln\left[\frac{H + \delta H^{1/2} + \delta^2}{(\delta - H^{1/2})^2}\cdot\frac{(\delta - 1)^2}{1 + \delta + \delta^2)}\right]. \tag{10.8}$$

Figure 10.4(a) shows Eq.(10.7) for $C > 1$. For $C > 2$, the effect of C is small. In Figure 10.4(b) the final drawdown level $Z(H = 0) = Z_\infty$ is plotted as a function of C^{-1}.

The temporal evolution of the function $z(t)$ can be determined with Eq.(10.2) when eliminating h with Eq.(10.6) to yield for $\beta = 1$

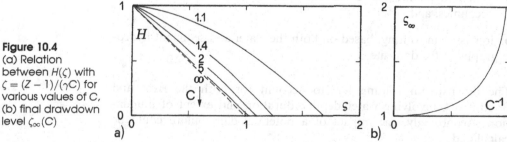

Figure 10.4
(a) Relation between $H(\zeta)$ with $\zeta = (Z - 1)/(\gamma C)$ for various values of C, (b) final drawdown level $\zeta_\infty(C)$

$$\alpha C_d(2g/h_o)^{1/2}t = -C^{1/2}\left[\frac{1}{2}\ln\left(\frac{C^{1/2}+H^{1/2}}{C^{1/2}-H^{1/2}}\cdot\frac{C^{1/2}-1}{C^{1/2}+1}\right)\right.$$

$$\left.+\arctan\left(\frac{(H^{1/2}-1)C^{1/2}}{H^{1/2}+C}\right)\right], \tag{10.9}$$

$$\text{and}\quad 3\alpha g C_d^2 t = -\ln\left[\frac{H^{3/2}(1-C^2)}{H^{3/2}-C^2}\right],\ \text{for}\ \beta=2. \tag{10.10}$$

To reach $H=0$, an infinite time is thus needed. Also, the time increases as C is decreased. For $C>2$, the effect of C is insignificant and $H=[1-(1/2)\alpha C_d(2g/h_o)^{1/2}t]^2$ for $\beta=1$, and $H=\exp(-2\alpha g C_d^2 t)$ for $b=2$, independent from both C and γ.

The present method was successfully applied by Singh (1996) to model the breach evolution of failed prototype dams. The reservoir surface was estimated as the reservoir volume divided by the reservoir filling height. The effect of the erosivity coefficient α is significant and both α and relative time vary linearly with C. From a series of historical cases it was deduced that α in the linear model ($\beta=1$) is about one order of magnitude larger than for the quadratic erosion model ($\beta=2$). The performance of the linear erosion rate was better than the quadratic erosion model, but there is a definite lack of data for the erosion process. From the available data, one may estimate $\alpha=0.15(H_s/H_d)^{-2}$ for $\beta=1$, and $\alpha=4\times10^{-3}(H_s/H_d)^{-1}$ for $\beta=2$, for dams that failed after 1920. The correlation for $\beta=2$ is better than for $\beta=1$.

According to Singh (1996), much experimental and numerical work has to be done in this field to understand the basic features of breaching modelling. His model can be considered the first semi-empirical model for the evolution of dam breaching. A complete breach model would include knowledge of:

- reservoir routing,

- breach dynamics as a function of hydrodynamics and soil mechanics, and

- downstream routing, based on both the water and sediment hydrographs at the dam site.

The downstream routing has to account for both the river and floodplains, involving water depths, duration and extent of inundation. Accordingly, the impact of a potential dam failure could be calculated.

10.3   DE SAINT-VENANT EQUATIONS

The equations of the dambreak wave have been derived by De Saint-Venant (1871). A 1D flow was assumed where the pressure distribution is hydrostatic and the velocity distribution uniform. With x as location measured from the dam section, t as time measured from the beginning of break, v as cross-sectional average velocity, h as flow depth, A as cross-section, S_o as bottom slope and S_f as friction slope (Figure 10.5), these equations are (Liggett, 1994, Chaudhry, 1993)

$$\frac{\partial A}{\partial t} + \frac{\partial(vA)}{\partial x} = 0, \tag{10.11}$$

$$\frac{1}{g}\frac{\partial v}{\partial t} + \frac{\partial}{\partial x}\left(h + \frac{v^2}{2g}\right) = S_o - S_f. \tag{10.12}$$

The first equation satisfies conservation of mass, i.e. is the continuity equation. The temporal change of cross-section plus the spatial change of discharge have to be always zero. The second dynamic equation requires that the temporal change of velocity plus the spatial change of energy head is always equal to the bottom slope minus the friction slope. With B as the free surface width the *celerity* c of a wave is defined as

$$c = (gA/B)^{1/2}. \tag{10.13}$$

The characteristic equations of the hyperbolic system (10.11), (10.12) have some interesting features and read

$$\frac{dv}{dt} \pm \frac{g}{c}\frac{dh}{dt} = g(S_o - S_f), \tag{10.14}$$

$$\frac{dx}{dt} = v \pm c. \tag{10.15}$$

First, the two sets are completely equivalent. Whereas the set (10.11; 10.12) is expressed as nonlinear partial differential equations, the set (10.14; 10.15) comprises two ordinary differential equations (compatibility equations) that are valid along the two characteristic curves defined in (10.15). Eqs.(10.15) plot two curves on the $x - t$ plane that are referred to as the positive $(+)$ and the negative $(-)$ *characteristics*,

Figure 10.5
Definition of variables in dambreak flow, (a) longitudinal and (b) transverse section.

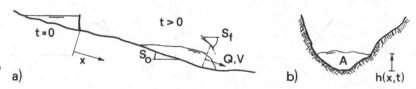

depending on the sign. For subcritical flow ($v < c$) the slope of the negative characteristic is negative, and the slope of the positive characteristic is positive. For supercritical flow ($v > c$) the slope of both characteristics is positive (Figure 10.6). Solving the set (10.11; 10.12) involves rectilinear coordinates and this is currently a standard procedure for finite difference methods (Chaudhry, 1993). The method of characteristics, in turn, has been popular in the sixties. The computation involves non-linear coordinates and thus difficulties in predicting locations where the unknowns $h(x,t)$ and $v(x,t)$ have to be computed. When shocks are generated, the set (10.14; 10.15) may be of particular interest, however.

The systems (10.11; 10.12) and (10.14; 10.15) are equations for the unknowns velocity v and flow depth h as functions of location x and time t. They are subject to appropriate *boundary and initial conditions*. Figure 10.7 shows the $x - t$ plane on which the two unknowns $v(x,t)$ and $h(x,t)$ are sought for the method of characteristics. A distinction between subcritical flow $v < c$ and supercritical flow $v > c$ has to be made. According to Cunge, et al. (1980), and with x_0, x_1 as upstream and downstream ends of the computational domain, one boundary condition has to be specified for each characteristic *entering* the computational domain at its limits. The initial condition may be considered as a special boundary condition in time. Therefore,

- two conditions have to be imposed as initial conditions at $t = t_0$,

- one boundary condition is needed both at the upstream ($x = x_0$) and downstream ($x = x_1$) boundaries for *subcritical* flow, and

- two boundary conditions are needed at the upstream boundary ($x = x_0$) for *supercritical* flow.

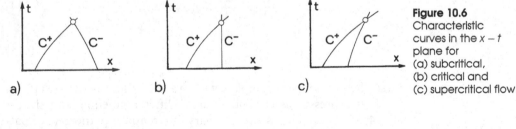

Figure 10.6
Characteristic curves in the $x - t$ plane for
(a) subcritical,
(b) critical and
(c) supercritical flow

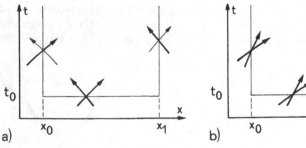

Figure 10.7
Characteristics at the limits of the computational domain, boundary or initial conditions in bold for
(a) subcritical and
(b) supercritical flow

The formulation of finite-difference methods is extensively discussed by, e.g. Chaudhry (1993). Particular aspects involve *computational stability* of various schemes, their *mathematical consistencies* and the treatment of initial and boundary conditions. This is not further treated in this chapter.

10.4 ONE-DIMENSIONAL DAMBREAK FLOW

10.4.1 Ritter solution

Although De Saint-Venant (1871) solved the equations he established for the particular case where the friction slope is compensated for by the bottom slope, i.e. $S_o - S_f \equiv 0$, for all x and t, Ritter in 1892 first used these equations for dambreak flows in rectangular channels (Hager and Chervet, 1996). The approach was extended by Su and Barnes (1970) to cross-sections of triangular ($\lambda = 2$), parabolic ($\lambda = 3/2$) and rectangular ($\lambda = 1$) shape by introducing the shape factor $a^2 = A/(Bh) = \lambda^{-1/2}$. The wave celerity is thus $c = a(gh)^{1/2}$. The generalized characteristic equations then read with $w = (2/a) (gh)^{1/2}$

$$\frac{d(v \pm w)}{dt} \doteq 0, \quad \text{or} \quad (v \pm w) = \text{const. along} \tag{10.16}$$

$$\frac{dx}{dt} = v \pm c. \tag{10.17}$$

Because the water is initially at rest the solution of (10.16; 10.17) reads

$$(gh)^{1/2} = \left(\frac{a}{2 + a^2}\right) \left[\frac{2}{a}(gh_o)^{1/2} - \frac{x}{t}\right], \tag{10.18}$$

$$v = \left(\frac{2}{2 + a^2}\right) \left[a(gh_o)^{1/2} + \frac{x}{t}\right]. \tag{10.19}$$

Eqs.(10.18; 10.19) satisfy the Froude similarity law.

Introducing dimensionless quantities scaled to the length h_o, the time $(h_o/g)^{1/2}$, and the velocity $(gh_o)^{1/2}$, the solution may be expressed, with $C = a(h/h_o)^{1/2}$ and $V = v/(gh_o)^{1/2}$, as

$$C = \left(\frac{a}{2 + a^2}\right) \left[2 - a\frac{X}{T}\right], \tag{10.20}$$

$$V = \left(\frac{2}{2 + a^2}\right) \left[a + \frac{X}{T}\right]. \tag{10.21}$$

The functions C and V are shown in Figure 10.8 as $h/h_o = (aC)^2$ and $v/(gh_o)^{1/2}$ both as functions of $m = X/T$. The characteristic features of those two plots are the positive velocity of propagation $m_+(C = 0)$ $= 2/a$, i.e. $m_+ = 2$, $6^{1/2} = 2.45$, $2^{3/2} = 2.83$ for rectangular, parabolic and triangular cross-sections, respectively. In turn, the negative velocity of propagation is $m_-(C = a) = -a$, i.e. $m_- = -1 =, -(2/3)^{1/2}$, $-(1/2)^{1/2}$, respectively. For equal initial depth h_o, a dambreak wave has the largest front velocity in the triangular section, and the largest upstream velocity in the rectangular section.

At the dam section $m = 0$ ($x = 0$) the flow depth remains constant $h_0/h_o = [2/(2 + a^2)]^2$, i.e. $h_0/h_o = (2/3)^2 = 0.44$, $(3/4)^2 = 0.56$, $(4/5)^2 = 0.64$, respectively. The free surface shape is parabolic for all three cross-sections, rather steep in the upstream portion, and tending to zero flow depth at the wave front. For the velocities, differences are even smaller than for the free surface profile, as is noted from Figure 10.8(b). Note the linear velocity distribution between the fronts of the dambreak wave.

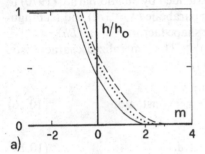

a)

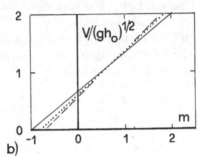

b)

Figure 10.8
Dambreak wave without friction for (—) rectangular, (. . .) parabolic, and (--) triangular horizontal channel (a) free surface, and (b) velocity as a function of $m = X/T = x$ /$[(gh_o)^{1/2}t]$ (Su and Barnes, 1970)

10.4.2 Dressler's asymptotic solution

Dressler (1952) investigated the effect of friction in the *rectangular channel* by setting for the friction slope according to Darcy-Weisbach

$$S_f = \frac{f}{4R_h}\frac{v^2}{2g}. \tag{10.22}$$

For a wide rectangular channel, the hydraulic radius is $R_h = h$, and $S_f = fv^2/(8gh)$. The governing system of equations is, from (10.11; 10.12) with $C = (h/h_o)^{1/2}$, $V = v/(gh_o)^{1/2}$, $X = x/h_o$, $T = (g/h_o)^{1/2}t$ and $R = f/8$

$$2\frac{\partial C}{\partial T} + C\frac{\partial V}{\partial X} + 2V\frac{\partial C}{\partial X} = 0, \tag{10.23}$$

$$\frac{\partial V}{\partial T} + V\frac{\partial V}{\partial X} + 2C\frac{\partial C}{\partial X} + R\frac{V^2}{C} = 0.$$

A perturbation solution for small values of RT was considered that can be expressed as

Figure 10.9
Dambreak wave
according to
Dressler (1952), (a)
relative flow depth
h/h_o and (b) relative
velocity
$\bar{U} = v/(gh_o)^{1/2}$ as
functions of m for
various σ.($\bullet$) wave
tip location

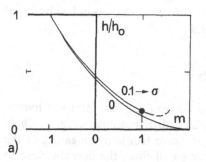

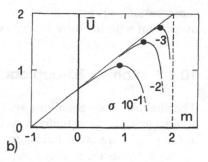

$$C = C^0 + C^1 TR + C^2 (TR)^2 + \ldots \qquad (10.25)$$

$$V = V^0 + V^1 TR + V^2 (TR)^2 + \ldots \qquad (10.26)$$

where C^0, V^0 are the zeroth-order solutions according to Ritter, as defined in (10.20; 10.21) for $a = 1$. With the perturbation parameter $\sigma = RT$, $M = (1 + m)/(2 - m)$ and when accounting for expansions up to order 1 the approximate solution reads

$$C = \frac{1}{3}(2 - m)\left[1 - \frac{1}{2}M^2 \sigma\right], \qquad (10.27)$$

$$V = \frac{2}{3}(1 + m)\left[1 - \frac{2M}{2 - m}\sigma\right]. \qquad (10.28)$$

For $\sigma = 0$, the original Ritter solution is obtained. It agrees with experiments up to about $m = 1$ (Dressler, 1952). At the front portion, the friction effect is significant, and the terms in square brackets have to be considered. It may also be noted that the friction causes a much greater effect on the velocity V than on the celerity C.

Figure 10.9 shows the relative flow depth h/h_o and the relative velocity $v/(gh_o)^{1/2}$ as functions of $m = X/T$ for various σ. For $\sigma = 0$ the solutions are as for the rectangular channel in Figure 10.8. For $\sigma > 0$, there is practically no change in the velocity for $m < 1$ (Figure 10.9(b)). For larger m, the curves split from the curve with $\sigma = 0$, much in analogy to a boundary layer. Dressler (1952) suggested that the solution for $\sigma > 0$ is valid up to the maximum velocity V_M, and introduced at the wave front (subscript F) the so-called wave tip of *constant velocity* $V_F = V_M$. From Eq.(10.27), the maximum (subscript M) of σ for any value of m is $\sigma_M = (2 - m)^2/(12M)$. Inverting gives approximately $m_M = 2 - 3\sigma^{1/3}$ and thus

$$V_M = 2(1 - \sigma^{1/3})[1 - \frac{2}{3}\sigma^{1/3}(1 - \sigma^{1/3})]. \qquad (10.29)$$

For $\sigma = 0.1$ as used in Figure 10.9(a) one has thus $C_M = 0.90$ and with $m_M = 0.61$, the corresponding flow depth is $h/h_o = 0.19$, from Eq.(10.27). The tip location is thus at $m = 1$. Dressler (1954) checked

these results against laboratory experiments and found substantial
agreement with the *wave tip approach*.

10.4.3 Pohle's 2D-approach

The shallow water equations are based on the assumptions of hydrostatic pressure and uniform velocity distributions. Clearly, these
assumptions are invalid close to the wave fronts propagating in the
up- and downstream directions. For small time, the domain close to
the dam section is thus governed by streamline curvature effects.

Pohle (1952), by using a *Lagrangian representation* for a particle
located at position x from the dam section at an elevation z above
the horizontal floor, sets

$$x = x_o + x^{(1)}(x_o, z_o)t + x^{(2)}(x_o, z_o)t^2 + \ldots \qquad (10.30)$$

$$z = z_o + z^{(1)}(x_o, z_o)t + z^{(2)}(x_o, z_o)t^2 + \ldots \qquad (10.31)$$

The functions $x^{(1)}; z^{(1)}$ represent velocities and $x^{(2)}; z^{(2)}$ are particle
accelerations. Because the water is initially at rest, $x^{(1)} = z^{(1)} = 0$.
Using *conformal mapping*, it is found that

$$x^{(2)}(x_o, z_o) = \frac{g}{2\pi} \ln \left[\frac{\cos^2\left(\frac{\pi z_o}{4h_o}\right) + \sinh^2\left(\frac{\pi x_o}{4h_o}\right)}{\sin^2\left(\frac{\pi z_o}{4h_o}\right) + \sinh^2\left(\frac{\pi x_o}{4h_o}\right)} \right], \qquad (10.32)$$

$$z^{(2)}(x_o, z_o) = -\frac{g}{\pi} \arctan \left[\frac{\sin\left(\frac{\pi z_o}{4h_o}\right)}{\sinh\left(\frac{\pi x_o}{2h_o}\right)} \right]. \qquad (10.33)$$

If expansions are carried to second order, Eqs.(10.30) to (10.33) define
the free surface profile for small times. For $z_o = h$ the result is $x^{(2)} = 0$
at the free surface, i.e. all particles move in the vertical direction after
an instance of the rupture. Upstream and downstream from the dam
section, the free surface is defined by Eq.(10.33)

$$\frac{h}{h_o} = 1 - \frac{gt^2}{\pi h_o} \arctan \left[\left(\sinh \frac{-\pi x}{2h_o} \right)^{-1} \right], \quad x \le 0 \qquad (10.34)$$

$$\frac{h}{h_o} = \frac{4}{\pi} \operatorname{arccot} \left[\exp\left(\frac{\pi x}{gt^2}\right) \right] - \frac{gt^2}{2h_o}, \quad x \ge 0. \qquad (10.35)$$

At the dam section ($x = 0$) one has simply

$$\frac{h}{h_o} = 1 - \frac{gt^2}{h_o}. \qquad (10.36)$$

Figure 10.10
Solution of Pohle
(1952) as modified
by Martin (1990) for
initial dambreak
wave ($T \leq 0.7$).($\bullet$)
Water depth at
dambreak section,
(o) positive and
negative wave
fronts.

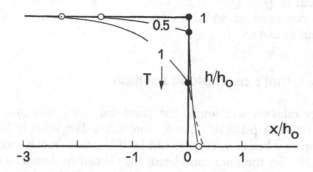

Accordingly, at the first instance, the particle located at the edge between the reservoir surface and the dam section moves like a free falling body. Later, the free surface is curved, in the reservoir with the centre of curvature below the free surface, and vice versa downstream from the dam section (Figure 10.10). Eqs.(10.34) and (10.35) may be approximated with $X = x/h_o$, and $T = (g/h_o)^{1/2}t$ as

$$\frac{h}{h_o} = 1 - \frac{1}{2}T^2[1 + \tanh X] \qquad X \leq 0, \tag{10.37}$$

$$\frac{h}{h_o} = 1 - \tanh(2X/T^2) - \frac{1}{2}T^2, \qquad X \geq 0. \tag{10.38}$$

The *front wave locations* $X_F(T)$ can be determined from Eqs.(10.37) and (10.38) when letting $h/h_o \to 0.99$ and 0, respectively, to read

$$X_F = \text{Arctanh}\left(\frac{0.02}{T^2} - 1\right) \qquad X_F < 0, \tag{10.39}$$

$$X_F = \frac{1}{2}T^2\text{Arctanh}(1 - \frac{1}{2}T^2). \qquad X_F \geq 0. \tag{10.40}$$

It is seen that $X_F \to 0$ for $T \to 2^{1/2}$. The validity of Eqs.(10.38) and (10.40) is thus restricted to $T << 1$, as was also demonstrated by Martin (1990). For very small times T, Eq.(10.40) can be approximated as

$$X_F = \frac{1}{2}T^{3/2} \tag{10.41}$$

which is much smaller than $X_F = 2T$ according to Ritter.
Martin (1990) has presented a complete model for the dambreak wave in the reservoir domain ($x \leq 0$) by using Pohle's solution for $T < 0.7$, and a modified Ritter solution for larger times. This solution was found accurate until no wave reflection at an upstream wall occurs. The effects of bottom slope are not included. Interestingly, the discharge across the dambreak section is

$$Q_o/Q_R = T/0.7, \quad T < 0.7 \tag{10.42}$$

and equal to $Q_o = Q_R = (8/27)b(gh_o^3)^{1/2}$ as given by Ritter. The discharge decreases as soon as the negative wave has reached the upstream boundary.

10.4.4 Hunt's asymptotic solution

Another relevant solution to the dambreak wave was presented by Hunt (1984). A plane of slope S_o contains a dam initially filled with water up to a height h_o (Figure 10.11). If locations much downstream of the dam location are considered, then it can be demonstrated that the leading terms of the dynamic equation (10.12) are $S_o - S_f = 0$. With the Darcy-Weisbach equation

$$S_f = \frac{fV^2}{8gh} \tag{10.43}$$

and the continuity equation for the rectangular channel

$$\frac{\partial h}{\partial t} + \frac{\partial(Vh)}{\partial x} = 0 \tag{10.44}$$

a so-called *outer solution* may be determined. Eliminating the friction slope S_f and using scale parameters (with asterisk *) $h^*, V^*, L^*, t^* = L^*/V^*$ yields, with $U = V/V^*, \Psi = h/h^*, \chi = x/L^*$, and $\tau = t/t^*$

$$1 - \frac{U^2}{\Psi} = 0. \tag{10.45}$$

This is the *kinematic wave approximation*. The scalings h^* and V^* can be obtained by a method outlined by Hunt. Eliminating the velocity in Eq.(10.45) gives for $\Psi(\chi, \tau)$ a nonlinear partial differential equation of first order

$$\frac{3}{2}\Psi^{1/2}\frac{\partial\Psi}{\partial\chi} + \frac{\partial\Psi}{\partial\tau} = 0. \tag{10.46}$$

The characteristic equations corresponding to Eq.(10.46) are $d\Psi/d\tau = 0$ along $d\chi/d\tau = (3/2)^{1/2}$. It may be seen that this system of equations involves a *kinematic shock* (subscript s) $\chi = \chi_s(\tau)$ and that $\Psi = 0$ in front of this shock. Upstream from the shock

Figure 10.11
Kinematic wave approximation for dambreak wave. (a) Definition of parameters, (b) height of shock as a function of location according to Eq.(10.48)

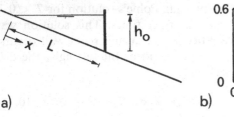

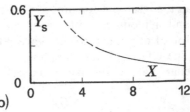

a)   b)

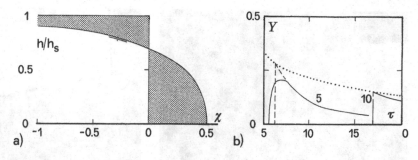

Figure 10.12
(a) Wave tip
according to
Eq.(10.50) and (b)
dambreak wave
according to Hunt
(1984). (. . .) shock
profile, (---) outer
solution, (—)
composite solution.

$$\Psi = (2\chi/3\tau)^2, \quad 0 < \chi \le \chi_s, \tag{10.47}$$

and at the shock, with $A = (1/2) \, L/h_o$ as the normalized reservoir volume

$$\Psi_s = \left(\frac{2A}{\tau}\right)^{2/3}, \text{or } \Psi_s = \frac{3}{2\chi}. \tag{10.48}$$

The solution is correct for locations $\chi = x/L \ge 5$ downstream of the dam (Hunt, 1984), where the origin of x is at the reservoir upstream end.

The *inner solution* of the problem relates to the shock mechanism. From an asymptotic analysis, Hunt is able to demonstrate that close to the shock, the velocities vary with τ but not with Ψ. Conservation of mass gives an additional relation to solve the complex system of governing equations. The location χ_s and free surface profile Ψ_s of the shock are

$$\chi_s = \frac{3}{2}\tau^{2/3} + \frac{1}{2}\tau^{-2/3}, \tag{10.49}$$

$$\frac{\chi - \chi_s}{\Psi_s} = \Psi + \ln(1 - \Psi) + \frac{1}{2}. \tag{10.50}$$

Figure 10.12(a) shows Eq.(10.50). The cross-hatched areas on both sides of the origin are equal, due to conservation of mass. The composite solution thus is equal to the sum of Eq.(10.49) and (10.50) minus (10.48) for $\chi < \chi_s$, and equal to Eq. (10.50) at the tip region (Figure 10.12(b)). Hunt stresses that his matched asymptotic expansion becomes asymptotically valid as soon as the shock has moved about five times the reservoir length.

10.5   EXPERIMENTAL APPROACH

10.5.1   Experimental observations

Lauber (1997) has conducted experiments in rectangular prismatic and hydraulically smooth channels. Based on preliminary experimentation, it was found that dambreak flows follow essentially the *Froude*

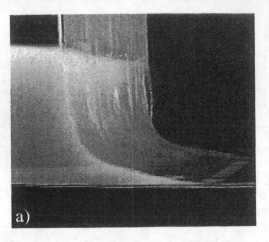
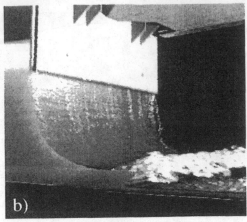

Figure 10.13 Wave fronts close to the dam section for initially (a) dry, and (b) wetted downstream channel

similarity law provided that the initial flow depth is at least 300 mm. Then, effects of surface tension and viscosity are negligible, at least up to 30 to 50 times h_o downstream of the dam section. Also, a dambreak may be considered instantaneous if the period of dam removal is smaller than $1.25(h_o/g)^{1/2}$. All experiments were conducted in the initially dry channel, and the effect of initial flow was found to be significant. Figure 10.13 compares the wave fronts for the dry and the initially wetted downstream channels, and it may be seen that the wave front is smooth for the dry, and of bore-type for the wetted channel. The wetting effect may be compared to flows that occur in hydraulically rough channels.

The data to be presented have been obtained with a high-speed video camera that was positioned successively at several sections along the 14 m long channel. Its width was 0.50 m, and the height 0.70 m. Observations were taken with a temporal interval of 50 ms at a specific section, and repetition of the dam break for the next section. A vertical gate without gate slots was used as the rupture mechanism, and a high degree of reproduction was obtained. Both flow depths and time-averaged velocities were measured to 1 mm, and to $\pm 0.05\,\mathrm{ms}^{-1}$. The bottom slope was varied from horizontal up to 50% (26.5°). This set of experiments is unique and was conducted both for the formulation of a semi-empirical approach, and the definition of a data basis for comparison with advanced numerical methods.

10.5.2 Horizontal smooth channel

Figure 10.14(a) shows a definition sketch of the configuration investigated. A dam of basin length L_B is initially filled with water up to a

Figure 10.14
Definition of
dambreak wave in
(a) horizontal
channel, (b) (—)
initial condition
$h(x, 0)$ and (---) free
surface after long
time

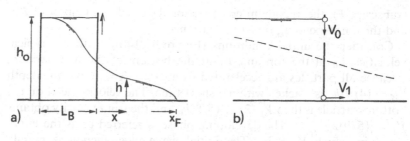

height h_o. The origin of the coordinate system $(x; h)$ is located at the
toe of the dam section. At time $t = 0$, the dam is suddenly removed and
the questions to be answered refer to the flow depth $h(x, t)$, the cross-
sectional average velocity $V(x, t)$, and the discharge $Q(x, t)$, where x is
the streamwise coordinate. Particular questions refer to the wave front

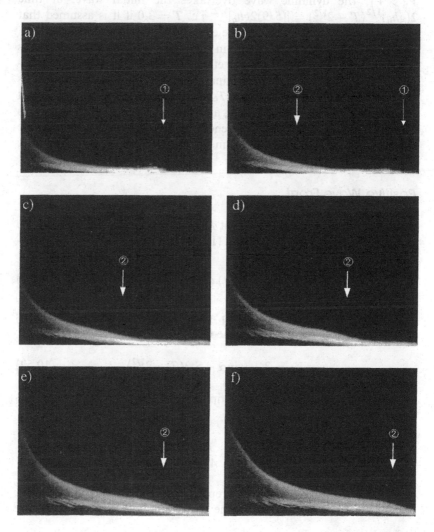

Figure 10.15
(a) Formation and
(b) overtaking of
the initial wave
① by the dynamic
wave ②

(subscript F), the maximum cross-sectional flow depth (subscript M), and the corresponding time of occurrence.

Consider the initial conditions (Figure 10.14(b)) with the typical velocities V_0 at the top and V_1 at the bottom of the dam section. Because all particles are accelerated from rest, $V_0 = gt$ and the depth $h_M = (4/9)h_o$ is reached within a short time. The velocity head on the bottom particle is then $V_1^2/2g = (5/9)h_o$ and the front velocity obtains $V_1 = [(5/9)2gh_o]^{1/2}$. The developing phase is referred to as the *initial wave*, in which $V_F = V_1$. The initial wave phenomenon is mainly governed by orifice flow at the dam section. The so-called *dynamic wave* with origin at the free surface of the dam section has a velocity $V_0 = 2(gh_o)^{1/2}$ according to Ritter and starts at time $T_0 = 2^{1/2}$, after having reached the channel bottom. Up to time T_0 this particle behaves essentially as a free falling body (Pohle, 1952). Because $V_0 > V_1$, the dynamic wave overtakes the initial wave, at time $2(gh_o)^{1/2}(T - 2^{1/2}) = [(5/9)2gh_o]^{1/2}$, i.e. $T = 3.0$ if it is assumed that the bottom slope compensates the effect of friction (Figure 10.15).

In the first instance, the flow in the vicinity of the dam section is governed by streamline curvature effects, and a 2D approach is appropriate (10.4.3). This phase is terminated when the dynamic wave starts at the dam section, i.e. at time $T_0 \cong 2^{1/2}$. Then all variables except those at the front vary gradually, and the De Saint-Venant equations apply. The time T_0 is not influenced by friction nor bottom slope because of the closeness to the dam section, and the short time period.

Positive Wave Front

The positive wave front (subscript F) in a rectangular channel may be determined from Eqs.(10.14) and (10.15)

$$\frac{d(v + 2c)}{dt} = g(\sin\alpha - S_f) \text{ along } \frac{dx}{dt} = v + c. \qquad (10.51)$$

Integrating the dimensionless Eq. (10.51) with the initial conditions $V_F(T = 2^{1/2}) = 2$ and $c = 0$ at the front gives

$$V_F = 2 + (\sin\alpha - S_f)(T - 2^{1/2}). \qquad (10.52)$$

The friction slope at the wave tip region, i.e. close to the front is equal to

$$S_f = \frac{f_a}{4h_a}\frac{V^2}{2g} \qquad (10.53)$$

where the hydraulic radius was replaced by the flow depth. Further, the friction coefficient f_a and the corresponding flow depth h_a are taken as averages $h_a = \sigma h_o$ and $f_a = 0.2 \mathbf{R}_a^{-0.2}$ for turbulent smooth flow. The average Reynolds number is $\mathbf{R}_a = 4q/\nu$ with $q = (8/27)$ $(gh_o^3)^{1/2}$ as the maximum discharge per unit width. For the experiments conducted with $h_o = 0.30\,\mathrm{m}$, $q = 0.251\,\mathrm{m^2 s^{-1}}$ and thus $\mathbf{R}_a = 0.61 \times 10^6$, the average friction factor is computed as $f_a = 0.014$. Inserting the friction slope $S_f = V_F^2 f_a/(8\sigma)$ in Eq.(10.52) and solving the quadratic equation for V_F yields, with $\tau = T - 2^{1/2}$ and $j = S_o \sigma/f_a$

$$V_F = \frac{4\sigma}{f_a \tau}\left[\left(1 + \frac{f_a}{\sigma}\tau(1 + \frac{1}{2}S_o \tau)\right)^{1/2} - 1\right]. \qquad (10.54)$$

Two cases are relevant:

- $\tau \ll 1$, for which $V_F = 2 + O(\tau)$, i.e. no effect of neither bottom nor friction slopes. The front velocity V_F then is equal to Ritter's prediction, and

- $\tau \gg 1$, for which to order τ^{-1}

$$V_F = 2(2j)^{1/2}[1 + \frac{1 - (2j)^{1/2}}{S_o \tau}]. \qquad (10.55)$$

Figure 10.16(a) compares observations of Lauber (1997) with Eq.(10.54) and substantial agreement is noted. All velocities start at the initial velocity $V_I = 1.05$, jump to the dynamic velocity $V_d = 2$ and follow the prediction for larger time. For all flows in both horizontal and sloping rectangular channels, the coefficient is $\sigma = 0.06$ and the tip region has typically a height of about 6% of the initial water depth h_o. For $j = 0.5$, corresponding to $S_o = j/4 = 0.125$ the front velocity remains constant and equal to 2. For large time, the front tends to the limit velocity $V_{F\infty} = 2(2j)^{1/2}$, as obtained from Eq.(10.55). It can

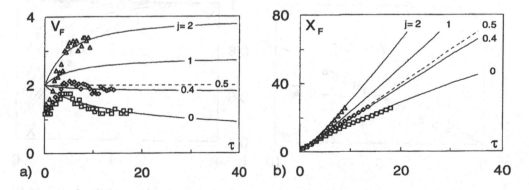

Figure 10.16 (a) Front velocity V_F for various bottom slopes S_o with $f_a/\sigma = 0.25$. (b) Front location $X_F(\tau)$ with $\tau = T - 2^{1/2}$. $S_o = $ ($\square$) 0, ($\lozenge$)0.1, ($\triangle$)0.5, (---) equilibrium slope.

be shown that $V_{F\infty}$ corresponds to the *uniform velocity* of the tip region. Pseudo-uniform flow conditions are practically reached when $\tau > 40$, for $j > 0.25$.

The wave front location $X_F(\tau)$ is determined by $dX_F/d\tau = V_F$ and imposing the initial condition $X_F(\tau = 0) = 0$. Integrating Eq.(10.54) leads to a singularity that cannot be removed. Therefore, the modified Eq.(10.55) has been used by imposing $V_F(\tau = 0) = 2$, i.e.

$$V_F = 2(2j)^{1/2}\left[1 + \frac{1 - (2j)^{1/2}}{S_o\tau + (2j)^{1/2}}\right] \qquad ,j > 0.1 \qquad (10.56)$$

$$V_F = \frac{4}{\bar{\tau}}[(1 + \bar{\tau})^{1/2} - 1] \qquad ,j = 0, \qquad (10.57)$$

where $\bar{\tau} = (f_a/\sigma)\tau$. Integration subject to $X_F(\tau = 0) = 0$ gives

$$X_F = 2(2j)^{1/2}\left[\tau + \frac{1 - (2j)^{1/2}}{S_o}\ln\left(1 + \frac{S_o\tau}{(2j)^{1/2}}\right)\right] \qquad ,j > 0.1 \qquad (10.58)$$

$$X_F = \frac{4\sigma}{f_a}[2(1 + \bar{\tau})^{1/2} - 2 + \ln\left[\left(4\frac{[1 + \bar{\tau}]^{1/2} - 1]}{\bar{\tau}[(1 + \bar{\tau})^{1/2} + 1]}\right)\right] \qquad ,j = 0. \quad (10.59)$$

Figure 10.16(b) compares observations with Eqs.(10.58) and (10.59) and agreement is noted. The positive wave front can thus be determined with an analytical approach.

Negative Wave Front

According to Ritter, the negative wave front (subscript nF) has a propagation velocity $V_{nF} = -1$. Figure 10.17(a) compares observa-

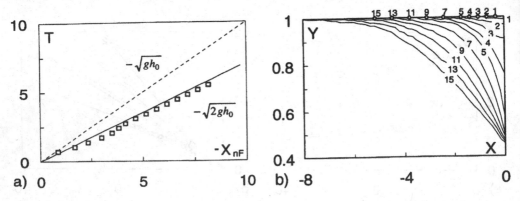

Figure 10.17 Negative wave front, (a) wave location $X_{nF}(T)$ according to (---) Ritter, (—) Eq.(10.61), (□) observations, (b) surface profiles at successive times $T = n \times \Delta T$ with $T = (g/h_o)^{1/2} \times 60\,\text{ms} = 0.343$, (◇) Eq.(10.61)

tions with this value, and considerable deviation is noted. According to Figure 10.17(b) that contains surface profiles in the reservoir reach, Ritter's solution is correct below the surface, but a surface current moves much faster than the wave body.

According to observations of Lauber (1997), the negative front velocity is $v_{nF} = -(2gh_o)^{1/2}$ or

$$V_{nF} = -2^{1/2}.\tag{10.60}$$

Integrating under the initial condition $X_{nF}(T = 0) = 0$ gives for the front location

$$X_{nF} = -2^{1/2}T, \quad T < 7.\tag{10.61}$$

Whereas the initial positive front velocity is smaller than predicted by Ritter, the negative front velocity is considerably larger. Ritter's solution is currently used for the *initiation of dambreak waves*. As a result the present initialization of dambreak waves is in contrast with detailed observations.

Maximum Flow Characteristics

At any location $x > 0$ the flow depth is initially $h = 0$ up to the arrival of the positive wave front, increases to a maximum h_M and decreases again. In the horizontal channel, the maximum flow depth is influenced by the relative location (x/h_o) and the relative basin length $\lambda_B = L_B/h_o$. Figure 10.18(a) shows a plot $Y_M = h_M/h_o$ as a function of the combined parameter $\bar{X} = \lambda_B(x/h_o)^{-2/3}$, and a perfect similarity may be noted. The data can be expressed, with notations barred, as (Lauber, 1997)

$$Y_M = \frac{4}{9}(1 + \bar{X}^{-1})^{-5/4}.\tag{10.62}$$

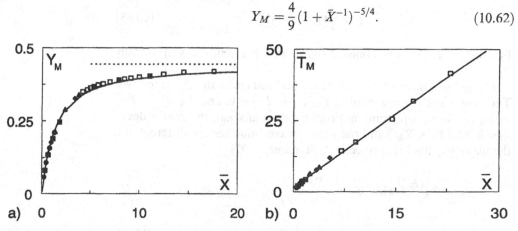

Figure 10.18 Characteristics of wave maximum (a) maximum flow depth $Y_M = h_M/h_o$ and (b) time of maximum $\bar{T}_M$ as functions of nondimensional location $\bar{X} = \lambda_B(x/h_o)^{-2/3}$.

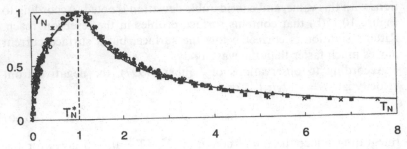

Figure 10.19
Generalized wave
profile $Y_N(T_N^*)$ for
$T_N^* < 1$ and $Y_N(T_N)$, for
$T_N > 1$. (–)
Eqs.(10.66), (10.67)

Interestingly, the maximum flow depth increases linearly with the relative basin length and decreases with $X^{2/3}$. The controlling parameter is $\bar{X} = L_B/(h_o x^2)^{1/3}$. The absolute maximum $Y_M = 4/9$ is identical with the Ritter solution and is asymptotically reached as $X \to 0$. For $X \gg 1$, the asymptotic solution of Eq.(10.62) is

$$h_M/h_o = (4/9)[L_B^3/(h_o x^2)]^{5/12}. \qquad (10.63)$$

The maximum flow depth is then directly related to the basin length L_B, and to the relative distance x/h_o from the dam.

The *time of maximum flow depth* $\bar{T}_M = (g/h_o)^{1/2} t_M (x/h_o)^{-2/3}$ increases linearly with $\bar{X} = \lambda_B X^{-2/3}$ as (Lauber, 1997)

$$\bar{T}_M = 1.7(1 + \bar{X}). \qquad (10.64)$$

This time corresponds to the propagation time of the negative wave upstream to the reflection boundary plus the propagation time to the section considered. At $X = 0$, this time is $\bar{T}_M = 1.7$, from Eq.(10.64).

The *propagation velocity* of the wave maximum is $d\bar{X}_M/d\bar{T} = 1.7^{-1}$ from Eq. (10.64). The corresponding Froude number $\mathbf{F}_M = (dX_M/dT) Y_M^{-1/2}$ is thus

$$\mathbf{F}_M = \frac{4}{3}(1 + X^{-1})^{5/8}. \qquad (10.65)$$

For large X, the so-defined Froude number tends asymptotically to $4/3$.

The general *wave profile* can be normalized (subscript N) as follows: The origin at the wave front is $Y_N(T_N = T_F) = 0$, and $Y_N(T_N = T_M) = Y_M$ at the wave maximum. From the data analysis, the time scales of the rising ($T_N < T_M$) and the falling wave branches are different and the wave profiles are, respectively (Lauber, 1997)

$$Y_N = T_N^{*1/3} \quad \text{with} \quad T_N^* = \frac{T - T_F}{T_M - T_F}, \quad 0 \le T_N^* < 1; \qquad (10.66)$$

$$Y = 1.1 T_N^{-1} \quad \text{with} \quad T_N = \frac{T}{T_M}, \quad T_N \ge 1. \qquad (10.67)$$

These two equations are valid for $\lambda_B \geq 3$, and allow determination of the complete wave profile downstream of the dam section for any time $T < 50$ and locations $X > 0$.

Wave Velocity

The time-averaged velocity $v(x,t)$ has been measured in the turbulent smooth flow with Particle Image Velocimetry (PIV) using vestiron grains of diameter 0.2mm. At a certain time and location, the transverse velocity distribution is nearly uniform, except for a thin boundary layer. Reynolds numbers upstream from the tip region are typically 10^5 to 10^6.

Figure 10.20 shows the dimensionless velocity $V(X)$, based on systematic experimentation for $\lambda_B = 11.05$. For $T < 22$, i.e. prior to wave reflection at the dam section, the velocity distribution is nearly linear, starting at $X = -X_F$, crossing $V = 2/3$ according to Ritter at $X = 0$, and extending to the positive wave front according to Eq.(10.59). For larger times, the velocity profiles turn about the endpoint $X = -\lambda_B$, are nearly straight up to a maximum and decrease gradually towards

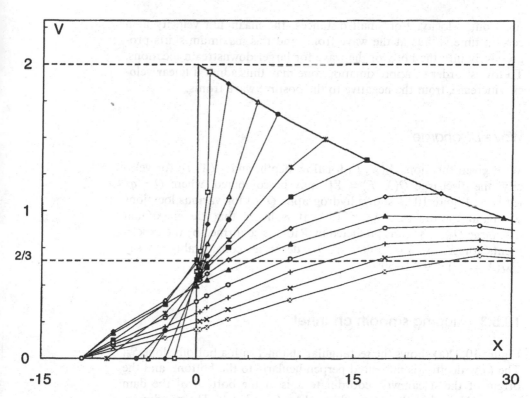

Figure 10.20 Velocity distribution $V(X)$ for various times $T=$(□) 2.05, (◆) 3, (△) 4.57, (●) 5.7, (*) 8.6, (■) 11.45, (◇) 17.15, (▲) 22.9, (○) 28.6, (+) 34.3, (x) 40.0, (⊡) 45.75. (–) Dynamic wave front, (---) Ritter velocity at dam section

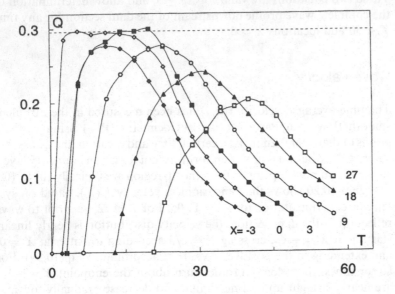

Figure 10.21
Discharge
distribution Q(T) for
various locations X,
with positive front
location and (---)
Ritter's solution
$Q_M = 8/27$.
$\lambda_B = 11.05$.

the front velocity. For small distances, the maximum velocity at a
certain time is thus at the wave front, and this maximum shifts progressively into the body of the wave for larger downstream locations.
To lowest order of approximation, one may thus admit a linear velocity increase, from the negative to the positive wave fronts.

Wave Discharge

With given functions $Y(X, T)$ for flow depth, and $V(X, T)$ for velocity, the discharge $Q(X, T) = YV$ may be computed, where $Q = q/$
$(gh_o^3)^{1/2}$. Figure 10.21 shows hydrographs $Q(T)$ for various locations
X, again for the case $\lambda_B = 11.05$. It is noted that the maximum
discharge $Q_M = 8/27$ as predicted by Ritter is confirmed by the experiments for $X = 0$. For larger X, Q_M decreases considerably, to say
$Q_M(X = 27) = 0.2 \ (-30\%)$.

10.5.3   Sloping smooth channel

Figure 10.22(a) shows the rectangular channel with a bottom slope S_o.
The flow depth h is measured perpendicularly to the bottom, and the
origin of the streamwise coordinate x is at the bottom of the dam
section. Note that the dam is limited to $L_B = h_o/S_o$. The nondimensional coordinates are $X = x/h_o$ for location, $T = (g/h_o)^{1/2}t$ for time
and $Y = h/(h_o\cos\alpha)$ for depth with α as the bottom inclination angle.

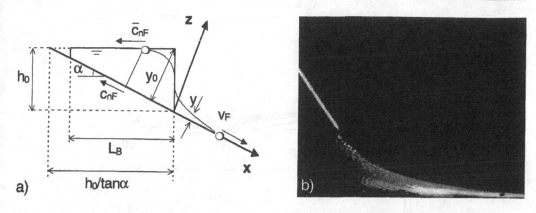

Figure 10.22 (a) Definition plot for dambreak wave in sloping channel, (b) Initial wave for $S_o = 0.5$ with channel bottom parallel to lower border of photo

Experiments were conducted for $S_o = 0.1$ and 0.5 and a generalized approach is presented.

Figure 10.23 shows the wave profiles $Y(X)$ for various times T and $S_o = 0.10$ and 0.50. For increasing slope, the downstream wave profile flattens, whereas a maximum flow depth occurs in the upstream region of the dam. For large times, the reservoir is emptied, and a compact wave with a moving positive and negative front flows down the channel.

The *positive wave front* has been determined for both the horizontal and sloping channel previously, and Figure 10.16 shows typical results. The *negative wave front* in the sloping channel with $L_B = h_o/S_o$ is given by $dx/dt = -[g(h_o + x\sin\alpha)]^{1/2}$. Integration under the initial condition $x(t = 0) = 0$ yields

$$T_R = \frac{2}{\sin\alpha}\left[1 - (1 - X\sin\alpha)^{1/2}\right]. \tag{10.68}$$

Figure 10.24 compares the modified Ritter solution (subscript R) with observations, and disagreement with observations is noted, as for the horizontal channel. Dividing T_R by $2^{1/2}$, i.e. $T_{nF} = T_R/2^{1/2}$, gives agreement as previously, such that

$$T_{nF} = \frac{2^{1/2}}{\sin\alpha}\left[1 - (1 - X\sin\alpha)^{1/2}\right]. \tag{10.69}$$

Solving for the wave front location gives

$$X_{nF} = -(\sin\alpha)^{-1}\left[1 - \left(1 - \frac{T\sin\alpha}{2^{1/2}}\right)^2\right]. \tag{10.70}$$

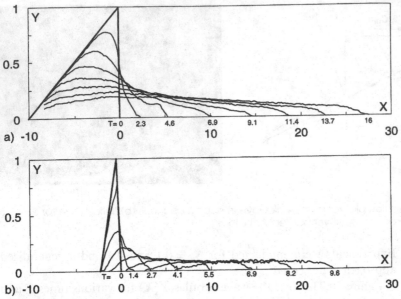

Figure 10.23
Wave profile $Y(X)$ for
various times T and
$S_o = $ (a) 0.10, (b)
0.50.

Lauber confirmed the validity of Eqs.(10.69) and (10.70) also for $S_o = 50\%$.

Wave maxima

For horizontal channels, the controlling parameter for the wave maximum is $\bar{X} = \lambda_B X^{-2/3}$. With $\varphi = (\sin\alpha)^{1/4}$ as the slope parameter, one may extend the dimensionless location to $\bar{X} = \lambda_B^{1-\varphi} X^{-2/3+0.2\varphi}$. Then, the maximum flow depths for the sloping channel follow Eq.(10.62)

$$Y_M = \frac{4}{9}(1 + \bar{X}^{-1})^{-5/4}. \qquad (10.71)$$

The agreement between observations and Eq.(10.71) is excellent for $\bar{X} > 0.2$. For large $\bar{X}$, i.e. large λ_B and small X, the maximum depth tends asymptotically to $Y_M = 4/9$, as predicted by Ritter. For small $\bar{X}$,

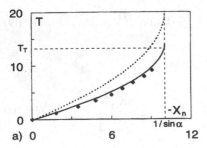

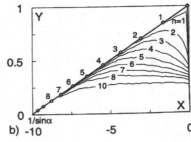

Figure 10.24
(a) Negative wave
front for $S_o = 10\%$,
(- - -) Eq.(10.68), (♦)
observations and
(—) Eq.(10.69). (b)
Negative wave
profiles $Y(X)$ for
various times
$T = n \times \Delta T$ with
$\Delta T = (g/h_0)^{1/2}$
$\times 0.2s = 1.14$, (◊)
front locations
according to
Eq.(10.70)

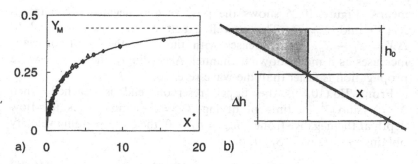

Figure 10.25
(a) Maximum flow depth $Y_M(X^*)$ with $X^* = h_o/(x\sin\alpha)$ for $S_o = (\diamond)$ 0.1, ($\circ$) 0.2, ($\triangle$) 0.5, (---) Ritter's solution. (b) Sketch of ratio $\Delta h = x\sin\alpha$ and h_o

i.e. large bottom slope, large distance or small reservoir length, the maximum flow depth tends to

$$Y_M = \frac{4}{9}\left[\frac{L_B}{h_o}\left(\frac{x}{h_o}\right)^{1/4}\right]^\varphi \left(\frac{h_o^5}{x^2 L_B^3}\right)^{5/12}. \qquad (10.72)$$

If the horizontal channel and bottom slopes $S_o < 0.05$ are excluded, and $L_B = h_o/S_o$, then $X^* = (h_o/x\sin\alpha)^{2/3}$ is a pertinent parameter. The maximum wave height is then simply (Figure 10.25)

$$Y_M = \frac{4}{9}[\tanh(0.067X^*)]^{1/3}. \qquad (10.73)$$

For small $X^* < 10$ one has $\tanh X^* \cong X^*$ and thus $Y_M = 0.18$ $(h_o/x\sin\alpha)^{2/9}$. The maximum flow depth thus decreases essentially with increasing location x/h_o and increasing angle α. The parameter X^* can be identified as the ratio of bottom elevation difference between the dam section and the point x, and the dam height. Accordingly, it does not matter whether x is large and $\sin\alpha$ small, but whether the product $x\sin\alpha$ is small or large compared to the initial flow depth h_o (Figure 10.25(b)).

The time of maximum flow depth occurrence T_M follows Eq.(10.64). With $\bar{T}_M = X^{-2/3}T$ and $\bar{X} = \lambda_B^{1-\varphi} X^{-2/3+0.2\varphi}$ it is as for the horizontal channel

$$\bar{T}_M = 1.7(1 + \bar{X}). \qquad (10.74)$$

This equation is thus applicable for any bottom slope up to 50%.

Emptying of channel

Whereas a horizontal channel empties only after a long time, the emptying of a sloping channel can be determined with elementary means. Figure 10.26 shows the positive and negative wave fronts $X_F(T)$ and $X_{nF}(T)$. The overall extent of the wave is equal to $\Delta X = X_F - X_{nF}$ and increases with time, i.e. the length of wave increases as it moves down a channel. Accordingly, the positive wave propagation is faster than the wave recession.

From Eq.(10.69), the upper reservoir end is reached when $X = -(\sin\alpha)^{-1}$, at time of drying $T_T = 2^{1/2}(\sin\alpha)^{-1}$. As the flow depth at the negative front is $h_{nF} = h_T = 0$, the corresponding velocity obtains $v_T = (\sin\alpha - S_f)gt$, or

$$V_T = (\sin\alpha - S_f)T. \qquad (10.75)$$

Inserting for $S_f = f_m V_T^2/(8\sigma)$ and solving for V_T gives

$$V_T = \frac{4\sigma}{f_a T}\left[\left(1 + \frac{f_a\sin\alpha}{2\sigma}T^2\right)^{1/2} - 1\right]. \qquad (10.76)$$

Further, with $dX_T/dT = V_T$ and the initial conditions $X_T(T = T_T) = -(\sin\alpha)^{-1}$, the drying front – time relation $X_T(T)$ is

$$\frac{f_a}{4\sigma}\left(X_T + \frac{1}{\sin\alpha}\right) = \left(1 + \frac{f_a\sin\alpha}{2\sigma}T^2\right)^{1/2} - \left(1 + \frac{f_a}{\sigma\sin\alpha}\right)^{1/2}$$

$$- \ln\left(\frac{1 + \left(1 + \frac{f_a\sin\alpha}{2\sigma}T^2\right)^{1/2}}{1 + \left(1 + \frac{f_a}{\sigma\sin\alpha}\right)^{1/2}}\right). \qquad (10.77)$$

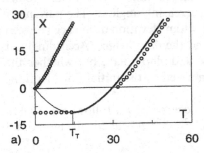

a)

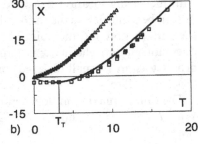

b)

Figure 10.26
Positive and negative wave fronts for $S_o = $ (a) 10%, (b) 50% for h_o[mm] $= (\triangle, \blacktriangle)$ 300 and 500 mm. $L_B = h_o/S_o$, (- -) length of wave

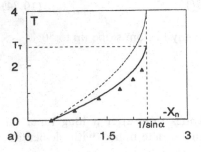

a)

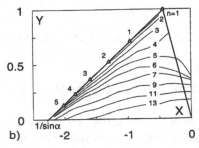

b)

Figure 10.27
(a) Negative wave front and (b) negative wave profiles for $S_o = 50\%$.

Figure 10.26 shows the negative front and the drying front, together with the positive front for both $S_o = 10\%$ and 50%. For small slopes the wave expands with increasing time, whereas it is of almost constant length for larger slopes. It is also noted that the prediction according to Eq.(10.77) for the drying front agrees with observations (Fig. 10.27).

Wave profiles

Figure 10.28 shows surface profiles $Y(T)$ for various locations X. For $S_o = 0.1$, the profiles have a single peak and the rising limb is much shorter than the falling limb. In contrast, waves in a steep channel have a plateau region of maximum depth and develop a nearly trapezoidal wave profile. In both cases, the maximum wave height decreases considerably with increasing location, as given in Eq.(10.73).

For small slopes a normalized wave profile can be determined, as for the horizontal channel. Normalizing with the time of positive front T_F, the time of negative front T_{nF}, the time of maximum flow depth T_M and the maximum flow depth h_M gives, with $Y_N = h/h_M$

$$Y_N = T_N^{*1/3}, \quad T_N^* = \frac{T - T_F}{T_M - T_F} \quad , \text{for } 0 < T_N < 1, \qquad (10.78)$$

$$Y_N = \exp[3(T_N - 1)^2], \quad T_N^* = \frac{T - T_M}{T_{nF} - T_M} \quad , \text{for } 1 < T_N < 2. \tag{10.79}$$

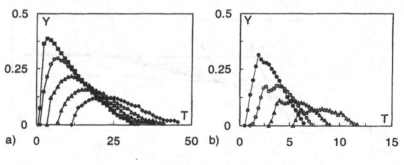

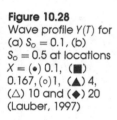

Figure 10.28
Wave profile $Y(T)$ for (a) $S_o = 0.1$, (b) $S_o = 0.5$ at locations $X = (\bullet)\,0.1$, ($\blacksquare$) 0.167, (o) 1, ($\blacktriangle$) 4, ($\triangle$) 10 and ($\blacklozenge$) 20 (Lauber, 1997)

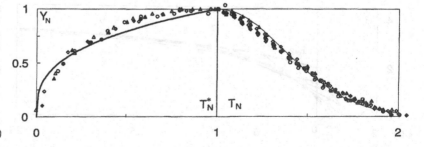

Figure 10.29
Generalized wave profile $Y_N(T_N^*)$, and $Y_N(T_N)$ for $S_o = 0.1$ and various X between 0.5 and 20

Eq.(10.78) for the rising wave profile is thus independent of the bottom slope.

Wave Velocity

The velocities in the body of the dambreak wave have been measured as in the horizontal channel. Experimental results are presented in Figure 10.30 as $V(X)$. The envelope curves of the positive and negative wave fronts are also plotted, and any velocity curve starts at the latter point $(X_{nF}; T)$, increases nearly linearly to a maximum velocity and tends to the positive front $X_F(T)$. Except for small times, the maximum velocity occurs somewhat upstream of the positive wave front. This effect must be attributed to the *wave tip region* with a strong effect of friction. As the wave depth increases, the velocity increases according to the dynamic equation. At the wave rear with a small flow depth, the velocity decreases again. For both $S_o = 0.1$ and 0.5, the maximum wave velocity is almost as large as the positive front velocity.

Wave Discharge

For given functions of flow depth $Y(X, T)$ and velocity $V(X, T)$, the discharge $Q = YV$ can be determined. Figure 10.31 relates to $S_o = 0.1$ and 0.5 again, and demonstrates that the wave body has a nearly constant discharge $Q = q/(gh_o^3)^{1/2} = 8/27$, according to Ritter for

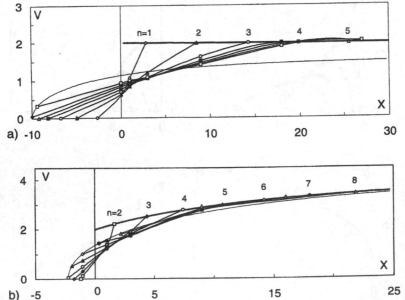

Figure 10.30
Wave velocity $V(X)$ at various times $T = n \times \Delta T$ with $\Delta T = (g/h_o)^{1/2}$ $0.5s = 2.86$ for $\lambda_B = (\sin\alpha)^{-1}$ and $S_o = $ (a) 0.1, (b) 0.5. (–) positive and (–) negative wave fronts

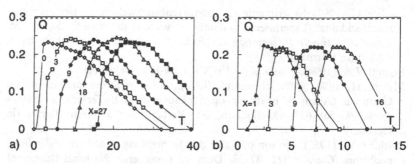

Figure 10.31
Discharge $Q(T)$ for
various locations X
and $\lambda_B = (\sin\alpha)^{-1}$ for
(a) $S_o = 0.1$, (b)
$S_o = 0.5$, (- - -)
$Q = Q_M = 8/27$.

$X = 0$. The present experiments indicate that $Q_M = 0.24$ is nearly independent of location and bottom slope, provided $X < 30$. Starting with $Q = 0$ at the negative wave front, the hydrograph increases rapidly to Q_M, and stays at the maximum discharge nearly all over the wave length. This result suggests that the effects of increasing flow depth and decreasing velocity in the wave body nearly compensate. In the wave body, the discharge is thus nearly constant and equal to the maximum discharge Q_M.

REFERENCES

Cassidy, J.J. (1993). Flood data and the effect on dam safety. Proc. Workshop *Dam Safety Evaluation*, Grindelwald **4**: 1–15.

Chaudhry, M.H. (1993). *Open-channel flow*. Prentice Hall, Englewood Cliffs, N.J.

Cunge, J., Holly, F.M., Verwey, A. (1980). *Practical aspects of computational river hydraulics*. Pitman, London.

De Saint-Venant, B. (1871). Théorie du mouvement non permanent des eaux, avec application aux crues de rivières et à l'introduction des marées dans leur lits (Theory of unsteady water flow, with application on floods in rivers and introduction of tides in their beds). *Comptes Rendues de l'Académie des Sciences* Paris **173**: 147–154; **173**: 237–240 (in French).

Dressler, R.F. (1952). Hydraulic resistance effects upon the dambreak functions. *Journal of Research* **49**(3): 217–225.

Dressler, R.F. (1954). Comparison of theories and experiments for the hydraulic dambreak wave. *Association Internationale d'Hydrologie* Rome **3**: 319–328.

Hager, W.H. and Chervet, A. (1996). Geschichte der Dammbruchwelle (History of dambreak wave). *Wasser Energie Luft* **88**(3/4): 49–54. (in German).

Hunt, B. (1984). Perturbation solution for dam-break floods. *Journal of Hydraulic Engineering* **110**(8): 1058–1071.

James, L.B. (1988). The failure of Malpasset dam. *Advanced dam engineering for design, construction and rehabilitation*: 17–27. R.B. Jansen, ed. Van Nostrand Reinhold, New York.

Jansen, R.B. (1988). *Advanced dam engineering for design, construction and rehabilitation*. Van Nostrand Reinhold, New York.

Lauber, G. (1997). Experimente zur Talsperrenbruchwelle im glatten geneigten Rechteckkanal (Experiments to dambreak wave in the smooth sloping rectangular channel). *Dissertation* **12115**. Eidg. Technische Hochschule ETH, Zürich (in German).

Liggett, J.A. (1994). *Fluid mechanics.* McGraw-Hill, New York.

Martin, H. (1990). Plötzlich veränderliche instationäre Strömunger in offenen Gerinnen (Abruptly unsteady open channel flows). *Technische Hydromechanik* 565–635, G. Bollrich, ed. Verlag für Bauwesen: Berlin (in German).

Pohle, F.V., (1952). Motion of water due to breaking of a dam, and related problems. *Circular* **521**: 47–53. Dept. of Commerce, National Bureau of Standards. Government Printing Office, Washington.

Schnitter, N.J. (1993). Dam failures due to overtopping. Proc. Workshop *Dam Safety Evaluation*, Grindelwald **1**: 13–19.

Singh, V.P. (1996). *Dam break modelling technology.* Kluwer Academic Publishers, Dordrecht.

Su, S.-T. and Barnes, A.H. (1970). Geometric and frictional effects on sudden releases. *Journal of Hydraulics Division* ASCE **96**(HY11): 2185–2200.

Subject Index

Author Index